AEG-TELEFUNKEN
FORSCHUNGSINSTITUT ULM
DOKUMENTATION

Nr. 10,260

MATERIALS PROCESSING – THEORY AND PRACTICES

VOLUME 4

Series editor

F.F.Y. WANG

NORTH-HOLLAND PHYSICS PUBLISHING
AMSTERDAM · OXFORD · NEW YORK · TOKYO

DRY ETCHING FOR MICROELECTRONICS

Edited by

RONALD A. POWELL
Solid State Laboratory
Varian Associates, Inc.
Palo Alto, California, USA

1984

NORTH-HOLLAND PHYSICS PUBLISHING
AMSTERDAM · OXFORD · NEW YORK · TOKYO

© Elsevier Science Publishers B.V., 1984

All rights reserved. No part of this publication may be reproduced, stored in a retrieval system, or transmitted, in any form or by any means, electronic, mechanical photocopying, recording or otherwise, without the prior permission of the copyright owner.

ISBN: 0 444 86905 0

Published by:

North-Holland Physics Publishing

a division of

Elsevier Science Publishers B.V.
P.O. Box 103
1000 AC Amsterdam
The Netherlands

Sole distributors for the U.S.A. and Canada:

Elsevier Science Publishing Company, Inc.

52 Vanderbilt Avenue
New York, N.Y. 10017
U.S.A.

Printed in The Netherlands

INTRODUCTION TO THE SERIES

Modern technological advances place demanding requirements for the designs and applications of materials. In many instances, the processing of materials becomes the critical step in the manufacturing processes. However, within the vast realm of technical literature, materials processing has not received its proper attention. It is primarily due to the lack of a proper communication forum. Since the materials processing is intimately concerned with specific products, those who are experts have no need to communicate. On the other hand, those who are involved with a different product will develop, in time, the technology of materials processing when required.

It is the objective of this series, Materials Processing – Theory and Practices, to promote the dissemination of technical information about the materials processing. It provides a broad prospective about the theory and practices concerning a particular process of material processing. A material process, intended for one technology, may have an applicability in another. This series provides a bridge between the materials engineering community and the processing engineering community. It is a proper forum of dialogues between the academic and the industrial communities.

Materials processing is a fast-moving field. Within the constraints of time and printed spaces, this series does not aim to be encyclopedic, and all-inclusive. Rather, it supplies an examination of material processes by the active workers. The view will be, by necessity, subjective. But the view will include both near-term and long-term prospectives. It is the fondest hope of this general editor that the volumes in this series can serve as first reference books in the field of materials processing.

Franklin F.Y. WANG
Stony Brook, New York

PREVIOUS VOLUMES IN THE SERIES

1. Fine line lithography
   R. Newman, volume editor
2. Impurity doping processes in silicon
   F.F.Y. Wang, volume editor
3. Laser materials processing
   M. Bass, volume editor

PREFACE TO VOLUME 4

Over the last decade, dry or plasma-assisted etching has emerged as an indispensable tool for the manufacture of microelectronic circuits, replacing traditional wet chemistry for etching and pattern transfer. The present volume collects together, for the first time, a series of in-depth, critical reviews of important topics in dry etching. This topical format provides the reader with more specialized information and references than found in a general review article. In addition, it presents a broad perspective which would otherwise have to be gained by familiarity with a large number of individual research papers.

As interest in dry etching has steadily increased over the years, so too has the number of technical papers. Hundreds have now been published including many excellent reviews of a general nature. Unfortunately, in-depth treatments of specific dry etch topics have been rare. Topical reviews are particularly lacking for areas which are promising but in an early stage of development, such as dry processing of III–V compound semiconductors, as well as areas which have been under intense development but are highly proprietary, such as dry etching of refractory metal silicides. In addition, there are topics which have been discussed more or less openly by workers for many years, such as dry etching aluminum and aluminum alloys, for which an article reviewing past experimental results in light of our current theoretical understanding of the dry etch process would be of great value.

The present volume arose out of the desire to provide, in one place, up-to-date critical reviews of these and other important topics in dry etching. The treatment of each specific topic in this volume is, to my knowledge, the most in-depth review of that topic yet published. An additional unique feature of the present book is the inclusion of an extensive literature review of dry processing, compiled by search of computerized data bases. Consisting of over 800 citations, this bibliography is intended to augment the more specific, topical references found throughout each of the critical review chapters. A subject index at the end of the book allows ready access to the key points raised in each of the chapters.

Chapters 1–3 deal with dry etching specific materials of proven or projected importance in microelectronics: aluminum and aluminum alloys (chapter 1), refractory metals and their silicides (chapter 2), and III–V compound semiconductors (chapter 3). In each case, successful etching depends on understanding a variety of interrelated materials and processing issues. Although a great deal of effort has been directed in the past at dry etching aluminum-based metallizations such as AlSi(2%) and AlSi(2%)Cu(1–4%), the development of a suitable, production-worthy etch strategy continues to challenge both process engineer and equipment manufacturer. Compared to aluminum etching, much less has been publically reported for refractory metals and metal silicides such as Mo, W, $TiSi_2$, WSi_2 etc. However, this is an area of intense and often proprietary development, since these materials are top contenders to replace or augment polysilicon as interconnects and metallizations in next-generation VLSI devices. Although the use of III–V compound semiconductors such as GaAs, $Al_xGa_{1-x}As$ and InP in electro-optics and discrete solid-state microwave devices has a long history, plasma-assisted etching of these materials is relatively unexplored. This situation is rapidly changing due to developments in integrated optics and digital GaAs technology which have placed more stringent demands on pattern transfer and etching of these materials.

Rather than dealing with a specific class of materials, chapter 4 critically reviews the dry etch technology of reactive ion beam etching (RIBE) – its instrumentation, basic principles and applications. Although the most common manifestations of dry etching today are reactive ion etching (RIE), plasma etching, and ion beam milling, recent developments in RIBE make it a promising alternative for certain microelectronic applications.

It is important to point out that the present volume deals only with the use of plasmas for the *removal* of materials, i.e., dry etching and pattern transfer. The use of plasmas to *deposit* thin films such as by plasma-enhanced chemical vapor deposition (PECVD) is not treated in either the critical review chapters or the bibliography.

Finally, I would like to thank the series editor, Prof. F.F.Y. Wang for his encouragement and the editorial staff of North-Holland Physics Publishing for the courteous and efficient manner in which they handled all aspects of this project. I also wish to acknowledge M.L. Manion, Stauffer Chemical Company, Mountain View, California, who compiled the subject index to this volume.

Ronald A. POWELL
Palo Alto, CA, USA
April, 1984

ADVISORY BOARD

Dr. David Dew-Hughes
   Department of Engineering Science, University of Oxford, Parks Road, Oxford, OX1 3PJ, UK
Dr. Stewart K. Kurtz
   Research and Development, Appliance Division, Clairol Inc., 83 Harvard Avenue, Stamford, Connecticut 06922, USA
Dr. S.A. Tobias
   Department of Mechanical Engineering, The University of Birmingham, South West Campus, P.O. Box 363, Birmingham B15 2TT, UK
Dr. John B. Wachtman, Jr.
   Center for Ceramics Research, Rutgers, The State University of New Jersey, Piscataway, New Jersey 08854, USA
Dr. J.H. Wernick
   Manager, Division of Materials Science Research, Bell Communications Research, 600 Mountain Avenue, Murray Hill, New Jersey 07974, USA
Dr. James Wei
   Department of Chemical Engineering, Massachusetts Institute of Technology, Cambridge, Massachusetts 02139, USA
Dr. A.F.W. Willoughby
   Department of Engineering Materials, The University, Southampton SO9 5NH, UK
Dr. S.M. Wu
   Department of Mechanical Engineering, University of Wisconsin-Madison, 1513 University Avenue, Madison, Wisconsin 53706, USA
Dr. T.S. Yen
   Academia Sinica, Beijing, The People's Republic of China

CONTENTS

Introduction to the series   v
Previous volumes in the series   vi
Preface to volume 4   vii
Advisory board   x
Contents   xi
1. D.W. Hess and R.H. Bruce
   Plasma-assisted etching of aluminum and aluminum alloys   1
2. T.P. Chow, A.N. Saxena, L.M. Ephrath and R.S. Bennett
   Plasma etching of refractory metals and metal silicides   39
3. R.H. Burton, R.A. Gottscho and G. Smolinsky
   Dry etching of Group III–Group V compound semiconductors   79
4. R.A. Powell and D.F. Downey
   Reactive ion beam etching   113
5. L.C. Molieri
   Dry etching for microelectronics – A bibliography   215
Subject index   295

CHAPTER 1

PLASMA-ASSISTED ETCHING OF ALUMINUM AND ALUMINUM ALLOYS

DENNIS W. HESS

Department of Chemical Engineering
University of California
Berkeley, California 94720, USA

RICHARD H. BRUCE

Xerox Corporation
Palo Alto Research Center
Palo Alto, California 94303, USA

Dry Etching for Microelectronics, edited by R.A. Powell
© *Elsevier Science Publishers B.V., 1984*

Contents

1. Introduction — 3
   1.1. Need for dry etching — 3
   1.2. Sputtering and ion beam etching — 4
2. Mechanistic considerations — 5
   2.1. Etchant production — 5
   2.2. Etching processes — 8
   2.3. Ion bombardment — 10
3. Etch chemistry of pure aluminum — 15
   3.1. Chlorine-based gases — 17
      3.1.1. Native oxide — 17
      3.1.2. Aluminum — 18
      3.1.3. Contaminants — 25
   3.2. Process requirements — 28
      3.2.1. High versus low pressure etching — 28
      3.2.2. Selectivity — 29
4. Aluminum alloy etching — 33
5. Corrosion — 33
6. Safety aspects — 35
References — 36

1. Introduction

Phenomenal advances in the design and fabrication of solid state devices and integrated circuits have occurred over the last two decades. Many of these advances have been made possible by the continual reduction in minimum feature size of individual circuit elements. Recent improvements in resist materials and exposure tools allow the printing of submicron patterns. Device fabrication, however, requires controllable transfer of these patterns into inorganic and organic film materials. Since liquid etching is limited to lateral dimensions on the order of 2 µm or greater, future, and indeed many current device designs, demand dry etching techniques.

Very large scale integration (VLSI) necessitates the interconnection of more than 10^5 devices per chip. Thus, a large portion of the chip area, more than 60% in some devices (Sinha, 1982), is taken up by interconnect layers. Currently, the most common metallization layer for electronic devices is aluminum or its alloys. Although pure aluminum presents a number of problems for VLSI (Learn, 1976), the most serious of these are electromigration and hillock formation. However, these problems can be alleviated by using aluminum–copper alloys. Thus, for the foreseeable future, aluminum or its alloys will certainly be used for at least one layer of metallization in integrated circuits.

In this chapter the chemical and physical processes occurring during the plasma-assisted etching of aluminum and its alloys are discussed. Since excellent reviews exist for sputtering of metal films (Vossen and Cuomo, 1978; Thornton and Penfold, 1978; Fraser, 1978; Waits, 1978), this topic will only be briefly mentioned here. The primary emphasis will be on mechanistic considerations and on effective use of the process latitude available for plasma-assisted aluminum etching.

1.1. Need for dry etching

Until the late 1970s, liquid etching was the preferred method of pattern delineation for thin films. Indeed, from a manufacturing standpoint, liquid etching provides low cost and often infinite selectivity (etch rate of the film being etched to the etch rate of the underlying film or substrate). Unfortunately, inherent limitations in solution etching preclude the use of this technique for micron and submicron pattern sizes. The most serious of these limitations is the isotropic etch characteristics, which results in undercutting of the mask material (and thus alteration of pattern dimensions). As a result, those dry etching techniques which can generate anisotropic etch profiles, have come into favor in recent years.

1.2. *Sputtering and ion beam etching*

In conventional sputtering and ion beam milling processes, chemically inert gas ions are accelerated to high energies ($> 100 \, \text{eV}$) and impinge on a substrate surface. When these ions strike the surface, they transfer their momentum to the material. If the ion energy is above a particular threshold value, substrate atoms, molecules, and ions are ejected; thus, etching is achieved. In sputtering, the material to be etched is immersed in a glow discharge where ions are accelerated across a dark space or sheath (Vossen and Cuomo, 1978). In ion beam etching (Hawkins, 1979; Lee, 1979), a confined plasma is used to generate ions. A set of grids used in the confinement process is biased so that an ion beam can be extracted from the source and directed onto a substrate surface.

Due to the highly directed nature of the ions used in the above processes, essentially vertical etch profiles are possible. Further, with ion beam milling, substrate tilting permits the generation of tapered profiles (Hawkins, 1979; Lee, 1979).

Unfortunately, sputter etching and ion beam milling have a number of shortcomings. Since both techniques use purely physical processes to remove substrate or film material, their selectivity is generally poor. For the most part, selectivity in these etch systems depends upon sputter yield differences between materials. Because sputter yields for most materials are within a factor of three of each other, selectivities are generally not adequate. In addition, since the ejected species are not inherently volatile, redeposition and trenching can occur (Mogab, 1980). Finally, the etch rates of physical processes such as those described above are inherently low.

All of the above-mentioned problems can in principle be avoided by adding a chemical component to the purely physical etching mode. Implementation of chemical/physical etching is generally performed by striking an r.f. glow discharge in a gas which decomposes to form species capable of reacting directly with the substrate to form a volatile chemical product. With aluminum, this can be accomplished by placing the aluminum-coated substrates within a glow discharge which contains chlorine-based gases. In this way, neutral species such as Cl_2 and Cl formed in the discharge can chemically react with aluminum to form volatile chlorides. Due to etch product volatility, and to the fact that neutral species, which are the primary etchants, are present in higher concentrations than are ions in r.f. glow discharges, an order of magnitude increase in etch rates over those observed with sputtering can be achieved. In addition, ion bombardment resulting from the difference in mobility between electrons and ions in the plasma atmosphere (see section 2.3) assists removal of native aluminum oxide, residues generated from the etch gas or photoresist and involatile alloy materials. Further, ion bombardment indirectly promotes anisotropic etching (see section 3.1.2).

2. Mechanistic considerations

By 1976, the feasibility of aluminum etching in chlorine-containing plasmas had been demonstrated (Poulsen et al., 1976). Although many studies have been carried out on aluminum plasma etching since that time, reproducible and controllable production-scale etching of aluminum and particularly its alloys has proved difficult. In order to understand the causes behind these difficulties and to establish viable etch procedures, it is necessary to examine the decomposition of chlorine-containing gases in r.f. glow discharges. Further, the interaction of the plasma-generated fragments with the aluminum surface must be delineated.

2.1. Etchant production

A number of different vapors have been used to etch aluminum in a glow discharge. However, since only CCl_4 has been extensively investigated with respect to the decomposition products formed in a plasma, the present discussion will concentrate on this material.

Exposure of an aluminum surface to CCl_4 liquid or vapor at room temperature results in little reaction (Minford et al., 1959). However, when a glow discharge is ignited in CCl_4 vapor, electron impact reactions create a number of product species, some of which are capable of rapidly reacting with aluminum and its native oxide. Interestingly, the number of major decomposition products is small; CCl_4, C_2Cl_6, Cl_2, and C_2Cl_4 account for more than 85% of the input CCl_4 (Tokunaga and Hess, 1980; Tiller et al., 1981). Thus a relatively simple reaction scheme can be proposed to account for the products generated (Tokunaga and Hess, 1980; Hess, 1982).

$$e + CCl_4 \rightarrow CCl_3 + Cl + e \tag{1a}$$

$$\rightarrow CCl_2 + 2\,Cl + e \tag{1b}$$

$$\rightarrow CCl_2 + Cl_2 + e, \tag{1c}$$

$$Cl + Cl \rightarrow Cl_2, \tag{2}$$

$$2CCl_3 \rightarrow C_2Cl_6, \tag{3}$$

$$2CCl_2 \rightarrow C_2Cl_4, \tag{4}$$

$$CCl_3 + Cl \rightarrow CCl_4, \tag{5}$$

$$CCl_2 + Cl \rightarrow CCl_3. \tag{6}$$

Of course, due to the large electron attachment cross-sections of chlorinated species (Buchel'nikova, 1959; Christophorou and Stockdale, 1968) and therefore to the rapid kinetics of low energy electron capture (Mothes et al., 1972), a number of negative ions are probably present in the CCl_4 plasma atmosphere. In fact, rapid dissociative electron attachment may be the primary dissociation mechanism of CCl_4 (Mothes et al., 1972; Tiller et al., 1981).

$$e + CCl_4 \rightarrow CCl_4^{*-} \rightarrow CCl_3 + Cl^-, \tag{7}$$

with the rapid kinetics of eq. (7) leading to the observed small number of dissociation products (Tiller et al., 1981).

At low pressures ($\sim 2\,Pa$), where electron energies are high, an even simpler reaction scheme has been proposed for CCl_4 decomposition (Tiller et al., 1981):

$$e + CCl_4 \rightarrow CCl + 3/2\,Cl_2 + e, \tag{8}$$

$$e + CCl \rightarrow C + Cl + e. \tag{9}$$

Proper choice of first-order rate constants for reactions (8) and (9) allows simulation of the observed time dependence of the CCl concentration (Tiller et al., 1981).

As power density is increased in CCl_4 plasmas, the fraction of CCl_4 dissociated increases. In addition, this enhanced conversion of CCl_4 is

Table 1
Effluent composition (wt%) downstream from CCl_4 plasma using anodized aluminum and stainless steel electrodes (Tokunaga and Hess, 1980).

Species	Anodized Aluminum	Stainless Steel
CCl_4	73.8	57.1
C_2Cl_6	14.3	13.9
Cl_2	8.6	6.8
C_2Cl_4	0.8	8.3

reported to result in a significant increase in the amount of Cl_2 produced (Bruce, 1981c; Tiller et al., 1981). Thus, reactions such as (1) and (2) probably dominate the decomposition as power levels rise. In any case, an increase in aluminum etch rate is observed at elevated power densities, which is consistent with increased Cl_2 concentrations if Cl_2 is a primary etchant species (see section 2.2) and etchant supply is rate limiting.

The yield of decomposition products in a CCl_4 plasma is also dependent upon the electrode material in a parallel plate reactor (Tokunaga and Hess, 1980). Table 1 shows the difference in product concentrations (relative to the input CCl_4) on two electrode materials, anodized aluminum and stainless steel. Whereas the yields of CCl_4, C_2Cl_6, and Cl_2 are within 30% on the different electrode materials, the unsaturated chlorocarbon, Cl_2Cl_4, is a factor of 10 higher on stainless steel than on anodized aluminum. Further, nearly 98% of the input CCl_4 is recovered as the four product species with anodized aluminum electrodes; only 84% is recovered with stainless steel. In the latter case, a polymer deposit is observed within the reaction chamber. Although the polymers formed during etching may appear deleterious, controlled formation of such residues is generally needed to assure anisotropic profiles in aluminum etching (see section 3.1.2.).

The above results are consistent with surface chemistry considerations on stainless steel versus anodized aluminum (Tokunaga and Hess, 1980). 'Active' electrode materials such as stainless steel or nickel react with chlorine atoms or molecules and enhance recombination of chlorocarbon fragments. Concentrations of unsaturated (polymer precursor) species such as C_2Cl_4 and its probable parent, CCl_2, are therefore high (table 1), and polymer films result. Indeed, polymer films form readily in C_2Cl_4 plasmas. With anodized aluminum electrodes, radical recombination is reduced (Nordine and Le Grange, 1976), and small concentrations of oxygen are released by the electrode to react with unsaturated chlorocarbons. The result of these two effects is a decrease in the concentration of unsaturated chlorocarbon species, and thus reduced polymerization.

Other experimental conditions can also lead to enhanced polymer formation via the mechanisms indicated above. For instance, large wafer loads ('loading effect'), a rapidly consumable electrode material (aluminum), and the addition of hydrogen or hydrogen-containing gases (e.g. $CHCl_3$; see section 3.1.2) to CCl_4 promote polymer formation. Again, etchant species such as Cl or Cl_2 are depleted, thereby increasing unsaturated gas phase species, and inducing polymer deposition.

Polymer formation in carbon-containing atmospheres can be minimized, even on 'active' electrode surfaces, by controlling the chemistry of the discharge (Hess, 1982). Generally, this is most easily accomplished by addition of Cl_2 to minimize the concentration of unsaturates. However, high concentrations of Cl_2 in such mixtures favor isotropic etch profiles.

Considerably less information has been reported on the electron impact dissociation of BCl_3. However, it is known that r.f. glow discharge dissociation of BCl_3 represents a high yield pathway for the formation of B_2Cl_4 (Wortik et al., 1949) suggesting that reactions similar to (1) and (7) may occur. In this way, aluminum etchant species such as Cl_2 and Cl can be formed.

2.2. Etching processes

In a manner similar to other chemical reactions, plasma etch processes can be ideally considered in terms of a number of primary steps, as indicated in table 2 (Mucha and Hess, 1983). If any of these individual processes does not occur, the overall etch sequence terminates. For instance, a reactive species (e.g., an atom) can rapidly react with a solid surface to form a product; however, unless the product has sufficient vapor pressure at the

Table 2
Primary steps in the etching process.

1. Generation of etchant species
2. Transport of etchant to surface
3. Adsorption of etchant onto surface
4. Reaction
5. Desorption of product from surface
6. Transport of product away from surface

etch temperature to desorb, or the desorption is promoted by ion bombardment, etching is prohibited.

Chlorine has been shown to readily adsorb onto a clean aluminum surface (Smith, 1972). Apparently, this adsorption occurs in two states: molecular physisorption, which is followed by dissociation, and chemisorption as atoms, with the fractional surface coverage of chemisorbed chlorine [i.e., in the form $(Al–Cl)_{ads}$] approximately equal to one. These results suggest that rapid dissociation of physisorbed Cl_2 occurs on a clean aluminum surface to form Al–Cl bonds. From a mechanistic standpoint, the reaction of chlorine with aluminum may thus be essentially concerted; i.e., adsorption and reaction may be virtually simultaneous, followed closely by product desorption. The implication of these conclusions is that either Cl_2 or Cl· should be a suitable etchant for aluminum. The primary etchant species is then determined by the relative concentrations of Cl_2 and Cl· in the plasma atmosphere. For instance, in etching mixtures where Cl_2 has been intentionally added, this species appears to be the dominant etchant. When pure vapors such as CCl_4, BCl_3, or $SiCl_4$ are used, however, chlorine atoms may play a significant role in the etch chemistry.

The type of mechanistic arguments presented above are only valid if a clean aluminum surface is available for etching. Oxide, polymer, or other contaminant residues will interfere with the purely chemical mechanism alluded to in the previous discussion. In these cases, ion bombardment may be essential to the removal of contaminants, and thus to uniform, reproducible aluminum etching.

High vacuum studies of aluminum etching in chlorine ion beams have generated insight into the species desorbed from the aluminum surface during etching (Smith and Bruce, 1982). Product peaks observed with a quadrupole mass spectrometer during etching corresponded to AlCl, $AlCl_2$, and $AlCl_3$. However, the appearance potentials of all these species coincided with those observed for the sublimation product of anhydrous $AlCl_3$. These results suggest that the etch product of aluminum in chlorine-containing gases could be $AlCl_3$, although Al_2Cl_6 may also be consistent with these observations. In addition, such conclusions indicate that atomic aluminum (Marcoux and Foo, 1981) and AlCl emissions (Curtis, 1980) arise from the decomposition of etch product in the glow discharge.

Unlike silicon, aluminum etch product volatility does not increase as the halogen in the etch gas is changed from bromine to chlorine to fluorine. In fact, a somewhat opposite trend is expected, as shown in fig. 1. Due to the low vapor pressure of AlF_3, fluorine-based etching of aluminum is not

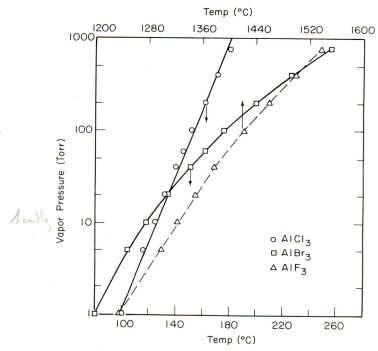

Fig. 1. Vapor pressure of AlF_3, $AlCl_3$, and $AlBr_3$, as a function of temperature. Data are from Stull (1947).

possible without a significant physical component (sputtering) in the etch process. However, formation of $AlCl_3$ or $AlBr_3$ allows chemical etching of aluminum. Indeed, consistent with the similarity of product vapor pressures, aluminum etch rates in BCl_3 and in BBr_3 are quite similar (Keaton and Hess, 1983).

2.3. Ion bombardment

Anisotropic etching is the result of a gasification reaction controlled in some way by bombardment of a directed species. In planar reactors the directed species are generally positive ions. Electrons and photons can also participate in such reactions, but to a substantially lesser extent than ions in most applications (Coburn and Winters, 1979). The directionality of the ion bombardment is a result of acceleration by intense electric fields

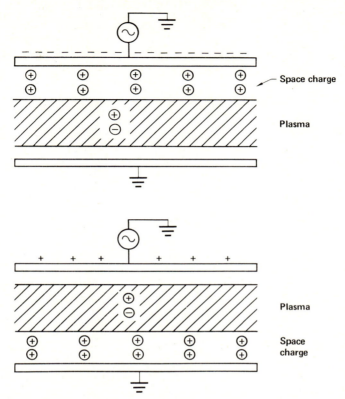

Fig. 2. Schematic of an r.f. glow discharge in a parallel plate etching system. The plasma region contains ions and electrons, while the sheath region contains only ions. The sheath region forms primarily near the most negative electrode.

in the vicinity of the electrodes. From the symmetry of a planar reactor, these fields are perpendicular to the electrode surfaces.

The role of the ion in the gasification reaction depends on both the material being etched and the etchant. In anisotropic etching of aluminum in chlorine-based etchant systems, the ions are thought to keep the etched surface free of species which would inhibit the spontaneous chemical reaction of chlorine and aluminum (Bruce and Malafsky, 1982). In silicon etching in chlorine or SiO_2 etching in fluorocarbon systems, the energetic ions are thought to directly enhance the reaction, possibly by damaging the surface to create sites more susceptible to etching (Flamm et al., 1983), or by promotion of product desorption (Coburn and Winters, 1979).

In a simplified view, a discharge in a planar system is composed of two regions (fig. 2). In the center of the discharge is the plasma, which is a highly conducting, relatively equipotential region containing both electrons and ions, with a charge density of about $1 \times 10^{10}\,\text{cm}^{-3}$. As a consequence of the high mobility of electrons relative to that of ions, the plasma region in r.f. discharges is maintained at an electric potential slightly above any of the surfaces in contact with the plasma. Any deviation forcing the plasma potential lower than that of its surroundings initiates an electron current from the plasma leaving a net positive accumulation of ions. The ions, due to their much lower mobility, cannot respond to the field as quickly as the electrons. The resulting net positive charge due to accumulated ions raises the plasma potential, which causes balanced electron and ion currents.

Since the plasma contains highly mobile electrons which shield the plasma from applied fields, it is relatively equipotential in comparison to the sheath regions. The applied electric fields therefore appear across the sheaths. The highly mobile electrons are quickly swept across the sheaths by these high fields leaving ions and neutrals in the sheath regions. Thus, space charge limited ion current flows across the sheath. At low pressures, in the absence of collisions in the sheath, the sheath thickness can be related to the ion current and voltage across it through the Langmuir–Child relationship. At higher pressures where ions experience many collisions traversing the sheath, the relation is more complicated.

This simplified view leads to a model of the discharge where the electric fields are confined to the sheath regions adjacent to the electrodes. The electric potential distribution for this arrangement is analogous to the series combination of two leaky capacitors. As shown in fig. 3, the sheath

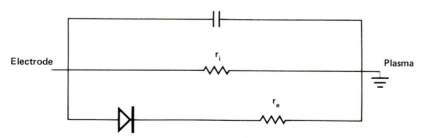

Fig. 3. Circuit diagram of a leaky capacitor used to describe the plasma sheath region. The diode simulates the large differences between electron and ion mobility. The impedance for electrons (r_e) is substantially less than the impedance for ions (r_i).

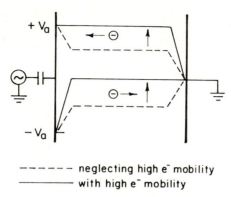

Fig. 4. Schematic of the potential distribution across a symmetrical parallel plate discharge (after Bruce, 1981a; reprinted with permission from Solid State Technology, Technical Publishing, a company of Dun and Bradstreet).

regions act as dielectrics and the plasma and electrodes act as conducting plates. The plasma, however, differs fundamentally from the electrodes in that the charges in the plasma are not strongly bound and will penetrate the sheath region. Because electrons respond more readily than ions to an applied field, the sheath region exhibits a diode-like behavior. Therefore, at r.f. excitation frequencies, the electron and ion currents from the plasma are balanced by the establishment of large electric fields across the sheath at the negative electrode which accelerate the low mobility ions from the plasma and by a small (if any) electric field across the positive electrode. In fact, the plasma can have an electric potential more positive than the positive electrode so that the electron current is controlled by thermal emission over the barrier. A schematic for the potential distribution for such a model is shown in fig. 4 for symmetrical (equal area) electrodes.

For situations where the electrodes are nonsymmetric, the potential distributions are different, particularly if the discharge is driven through a capacitively coupled network. In such a situation, neglecting the mobility of the charges in the plasma, the potential drop across the sheath adjacent to the small electrode will be larger than the drop across the sheath at the larger electrode (fig. 5). The initial electron current to the smaller electrode will therefore be much larger than the current to the larger electrode. In contrast to the electron current, the ion current depends very little on the sheath potentials due to space charge formation; consequently, the net flow to the electrode will depend primarily on the electrode area with the

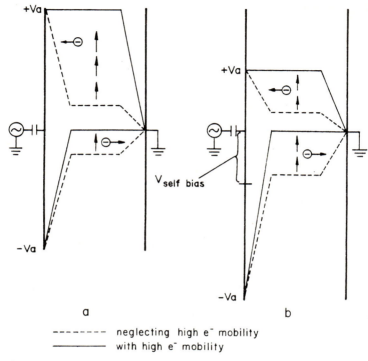

Fig. 5. Schematic of the potential across a capacitively coupled nonsymmetric parallel plate discharge: (a) potential distribution with net flow of charge to electrodes (transient condition); (b) potential distribution for no net charge flow to electrodes due to the presence of a self-bias voltage (steady-state condition). (After Bruce, 1981a; reprinted with permission from Solid State Technology, Technical Publishing, a company of Dun and Bradstreet.)

small electrode receiving less net ion current than the large electrode. The negative electrode will, therefore, accumulate negative charge until its potential is lowered sufficiently so that the ion and electron currents are balanced (fig. 5). The negative bias on the smaller electrode, known as the self bias, increases the ion bombardment energy at this electrode. The self bias is reduced as the chamber pressure is increased, particularly with electronegative gases such as chlorine (Horiike et al., 1982).

As described above, the ions are accelerated across the sheath regions toward the electrodes by inherent electric fields. The bombardment energy will depend on a number of discharge conditions such as pressure, excitation frequency, reactor geometry, operating voltage and ion mass. At excitation frequencies of 13.56 MHz, where most commercial aluminum etching equipment operates, the transit time for ions crossing the sheath is several periods of the driving oscillator (Bruce et al., 1981). Thus, the ions can have a maximum energy (in the absence of collisions in the sheath) equal to an average of the maximum electric potential appearing across the sheath. In the case of a symmetric electrode configuration, this average energy is equal to the applied voltage amplitude divided by π, while for the asymmetric configuration, the maximum bombardment energy at the smaller electrode is the applied voltage amplitude. The operating voltage is generally proportional to the r.f. power driving the discharge.

Particularly at high pressures (in the 100 Pa range), the ions experience collisions crossing the sheath which can substantially reduce the bombardment energy. In many cases the dominant collision process for energy reduction is the charge exchange reaction where the ion receives an electron from a neutral, transforming the neutral into an ion with very little energy. These reactions have high cross-sections (about ten times higher than any other collision cross-sections) for ions and neutrals of the same species. At 1 Pa, where most etching is done in asymmetric systems, most of the ions cross the sheath with little energy reduction (Coburn and Kay, 1972). At 100 Pa, where high rate etching is accomplished, however, a substantial reduction in the bombardment energy can occur (Bruce, 1981a). At higher pressures, even up to 1000 Pa, changes in the ion direction due to sheath collisions are small and practically unimportant for anisotropic etching of thin film materials (Bruce et al., 1981).

3. Etch chemistry of pure aluminum

Due to the thin layer ($\sim$ 3 nm) of native oxide always present on air-exposed aluminum films, and to the high reactivity of aluminum surfaces to water vapor or oxygen, controllable and reproducible plasma etching of aluminum requires that two processes occur prior to the start of aluminum etching (Tokunaga et al., 1981). These processes are the scavenging or reaction of plasma-generated species with water vapor and/or oxygen, and the removal of the native aluminum oxide layer. Collectively, these steps comprise what has been called an initiation or inhibition period, since a lag time is generally observed before commencement of aluminum etching. This effect can be seen in fig. 6, where the intensity of the 396.4 nm atomic emission line of aluminum is monitored as a function of etch time during

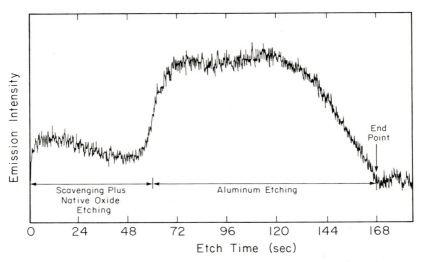

Fig. 6. Optical emission intensity of the 396.4 nm atomic emission line of aluminum, as a function of etch time. The etchant gas was pure CCl_4 at a pressure of 13 Pa, and a frequency of 13.56 MHz.

etching in a CCl_4 plasma. For the first minute after initiation of the glow discharge, only a weak aluminum line is observed due to the slow etching aluminum oxide layer. During this period, oxygen and water vapor are scavenged by the etch gas, the aluminum surface (thereby inhibiting etching) or both. The subsequent rise in emission intensity signifies the start of aluminum etching. The more intense emission continues until the aluminum begins to clear at the edges of the wafer (~ 60 s after aluminum etching begins). As the aluminum etches from the wafer edge to the center ('bull's eye' effect), the emission intensity falls until it reaches the initial level. At this point, etching is complete.

A sometimes overlooked fact is that both processes comprising the initiation period must be controlled if reproducible aluminum etching is to be achieved (Hess, 1982). Naturally, depending upon the etch gases and the plasma conditions used, either process can control the initiation time. Further, if oxygen and water vapor are effectively excluded from the chamber, and if low pressures or high energy ions are used to rapidly remove the native oxide, the initiation period can be negligible with respect to the total etch time. Thus it may appear that no initiation period exists.

Currently, chlorinated etch gases are the preferred etchants for aluminum; therefore, aluminum trichloride (AlCl$_3$) is the probable etch product. The relatively low vapor pressure of this material results in redeposition or adsorption on the etched sidewalls, the resist, and the chamber walls. Since AlCl$_3$ is a Lewis acid, rapid chemical reactions with water vapor, resists, and pump oils can occur, leading to etch reproducibility problems, corrosion, resist degradation, and safety concerns. In order to reduce the AlCl$_3$ left on reactor and substrate surfaces, the electrodes and/or the chamber are often intentionally heated (40–60°C) during the etch process to promote removal of AlCl$_3$ from surfaces within the etch chamber.

3.1. Chlorine-based gases

A number of chlorinated gases and gas mixtures have been used to etch aluminum. These include CCl$_4$, CHCl$_3$, BCl$_3$, PCl$_3$, SiCl$_4$, and their mixtures with each other, with Cl$_2$ and with 'inerts' such as Ar, He, and N$_2$. Further, these gases have been used in both low pressure asymmetric reactors (RIE) and in high pressure symmetric reactors. Regardless of the exact gas or mixture used, successful aluminum etching requires that the gases effectively scavenge water vapor and/or oxygen, uniformly and rapidly remove native aluminum oxide, do not form significant amounts of polymers or residues, show reasonable selectivity for resist materials, display etch rates above 0.1 µm/min, and are capable of anisotropic etching. To date, no single etch gas has been able to meet all these criteria. Thus, considerable efforts have been expended to develop suitable gas mixtures for aluminum etching.

3.1.1. Native oxide
Glow discharges of pure Cl$_2$, Br$_2$, or HCl cannot effectively etch aluminum prior to native oxide removal (Poulsen et al., 1976; P.M. Schaible et al., 1978). If, however, the native oxide is removed by sputtering, aluminum etches rapidly in Cl$_2$ with or without a plasma (Poulsen et al., 1976; Bruce, 1981a). These observations are consistent with the high strength of aluminum–oxygen bonds (~163 kcal/mole) (Vedeneyev et al., 1966), and with the mechanistic etch considerations proposed for pure aluminum in section 2.2.

As a result of the inability of pure chlorine to etch native aluminum oxide unless a significant amount of ion bombardment is present (Schaible et al., 1978; Heiman et al., 1980), gases such as CCl_4, BCl_3, or $SiCl_4$ are generally employed for aluminum etching. Any of these gases are capable of etching the native oxide layer, presumably by supplying high energy ion bombardment and/or by assisting in reduction of the oxide with CCl_x, BCl_x, or $SiCl_x$ species.

It has been reported (Tokunaga et al., 1981) that r.f. glow discharges of pure CCl_4 etch sputtered Al_2O_3 (and thus by inference, native aluminum oxide) more rapidly than do discharges of BCl_3. Other studies that used optical emission spectroscopy to monitor the induction period in aluminum etching (Spencer and Shu, 1982) indicate that the native oxide etches faster in BCl_3 than in CCl_4 discharges. These discrepancies appear due to differences in experimental conditions. First, native aluminum oxide is not Al_2O_3, so some differences in etch rate are to be expected. Second, the optical emission studies were performed in discharges containing 1–2% BCl_3 or CCl_4 in argon; therefore, the energetics of the plasma are quite different than in pure discharges of BCl_3 or CCl_4. Further, the results which indicate a higher native oxide etch rate in BCl_3 (Spencer and Shu, 1982), used anodized electrodes and an anodized etch chamber. This type of system could result in a modification of the chemistry relative to non-anodized chambers because of additional oxygen released from the walls (Tokunaga et al., 1981). In any case, differences in the etch rates of Al_2O_3 or of native aluminum oxide layers in BCl_3 and CCl_4 are most likely related to the relative concentrations and reactivities of plasma generated etchant species and to the relative plasma and electrode potentials which determine ion bombardment energies.

3.1.2. *Aluminum*

As previously mentioned, the reaction of clean aluminum with pure Cl_2 is spontaneous. This reaction has been observed not to be enhanced either by ion bombardment or by radical formation (Smith and Bruce, 1982), in contrast to other materials such as Si and SiO_2 (fig. 7). This is consistent with the insensitivity of the aluminum etch rate in Cl_2 to r.f. power in a planar reactor (Bruce and Malafsky, 1982) (fig. 8). Since an increase in r.f. power is known to increase the density of both excited states and radicals of Cl_2, the absence of an increase in etch rate with increased r.f. power strongly suggests that the principal reactant is the Cl_2 feed gas.

With a spontaneous etch reaction, anisotropic etching in pure Cl_2 is not possible unless additional species are mixed with Cl_2 to prevent reaction from occurring on sidewalls and undercutting the masking layer. These

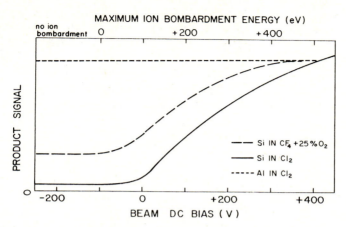

Fig. 7. Dependence of etch rate on ion bombardment energy for a beam of ions and radicals ejected from a confined plasma, and directed onto a sample surface positioned in a high vacuum chamber where etching occurs. (After Smith and Bruce, 1982; reprinted by permission of the publisher, The Electrochemical Society, Inc. This figure was originally presented at the Fall 1981 Meeting of the Electrochemical Society, held in Denver, Colorado.)

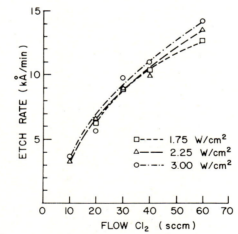

Fig. 8. Aluminum etch rate as a function of Cl_2 flow for various r.f. power levels. Other process parameters are: 20 sccm BCl_3, 250 sccm He, 133 Pa total pressure, 2.5 cm electrode spacing, and a frequency of 13.56 MHz. (After Bruce and Malafsky, 1982; reprinted by permission of the publisher, The Electrochemical Society, Inc. This figure was originally presented at the Fall 1981 Meeting of the Electrochemical Society, held in Denver, Colorado.)

additives or the plasma reaction products resulting from the gas mixture must inhibit etching at the sidewall of the feature, but not at the bottom. Such a situation is possible if the additives form a reaction–inhibiting layer on the sidewalls which is cleared by ion bombardment where etching is to occur. Since ions impinge perpendicular to the feature to be etched, anisotropic profiles can result from inhibiting species present on film sidewalls.

Several additives to Cl_2 which have been useful in inhibiting sidewall etching include: CCl_4, $CHCl_3$, CH_3Cl and $SiCl_4$. Similar results have been reported in both the low pressure RIE mode and the higher pressure regime (Horiike et al., 1982; Bruce and Malafsky, 1982). In the high pressure case, the degree of anisotropy was found to depend strongly on two process conditions: the gas mixture and the r.f. power. Increases in either the r.f. power (fig. 9) or in the relative amount of additive (fig. 10) increased the degree of anisotropy. The dependence of anisotropy passiva-

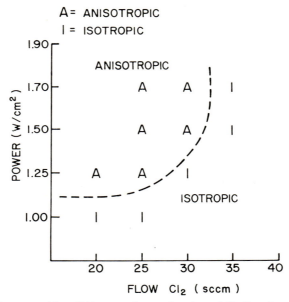

Fig. 9. Anisotropy with a 50% overetch as a function of Cl_2 flow for various r.f. power levels. Other process parameters are: 30 sccm BCl_3, 250 sccm He, 13.5 sccm $CHCl_3$, 133 Pa total pressure, 2.5 cm electrode spacing, and 13.56 MHz frequency. (After Bruce and Malafsky, 1982; reprinted by permission of the publisher, The Electrochemical Society, Inc. This figure was originally presented at the Fall 1981 Meeting of the Electrochemical Society, held in Denver, Colorado.)

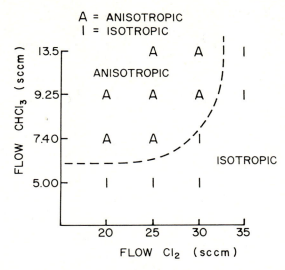

Fig. 10. Anisotropy with 50% overetch as a function of Cl_2 and $CHCl_3$ flow. Other process parameters are: 30 sccm BCl_3, 250 sccm He, 1.75 W/cm² at 13.45 MHz, 133 Pa total pressure, and 2.5 cm electrode spacing. (After Bruce and Malafsky, 1982; reprinted by permission of the publisher, The Electrochemical Society, Inc. This figure was originally presented at the Fall 1981 Meeting of the Electrochemical Society, held in Denver, Colorado.)

tion on power suggests that a discharge product, not the additive, is the inhibiting species.

In the RIE mode, anisotropy is observed to depend strongly on the gas mixture (fig. 11) and on the discharge pressure (fig. 12). Again, increases in inhibitor concentration increase anisotropy. Further, an increase in pressure is observed to decrease anisotropy. Since the ion bombardment energy is also reduced as pressure increases, the sidewall inhibition may be due to a redeposited material from the ion bombardment. It is also possible that the increased ion bombardment accelerates photoresist erosion, the fragments of which react with the additive to produce inhibitor. In addition, the concentration of radical species which attack the reaction–inhibiting layer could also increase with pressure.

The same molecules that have been added to Cl_2 plasmas to inhibit undercutting (CCl_4, $CHCl_3$, etc.) have also been used to etch aluminum by themselves. Since a major product formed in a discharge of these molecules is chlorine (Bruce, 1981c), etching could proceed as it would in

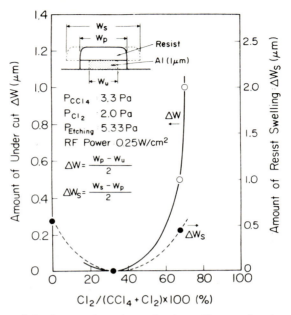

Fig. 11. Amount of aluminum undercutting and resist swelling as a function of Cl_2 added to CCl_4 (from Horiike et al., 1982; courtesy of Japan J. Appl. Phys.)

a mixture with Cl_2 gas but with a lower concentration of Cl_2. It has not yet been possible, however, to separate the etching contribution of the radicals formed in the discharge from that of Cl_2. The lower etch rates reported without the addition of Cl_2 reflect the reduction of this etchant. The inhibiting products should be more abundant in discharges without Cl_2 addition, than in the mixture case. In fact, many of these molecules show a strong propensity for polymerization on many surfaces, including reactor walls. It is thought that this polymerization is responsible for protecting the sidewalls from etching.

One of the first molecules reported useful in anisotropic plasma etching of aluminum is CCl_4 (Poulsen et al., 1976; Schaible et al., 1978). The etching is anisotropic but there are some difficulties reported with this molecule (Sato and Nakamura, 1982a). These difficulties include poor control of undercutting with overetching under conditions of high selectivity, contamination of the wafer and reactor with carbonaceous polymer, and deterioration of resists and the vacuum pump oil. The process can be

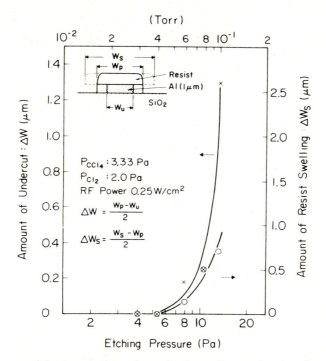

Fig. 12. Amount of aluminum undercutting and resist swelling as a function of total pressure (from Horiike et al., 1982; courtesy of Jap. J. Appl. Phys.).

improved somewhat through the addition of nitrogen (Sato and Nakamura, 1982b). The nitrogen reacts with carbon fragmented from CCl_4, thereby forming CN. This chemical removal of carbon reduces the formation of chlorocarbon polymer. In addition, since carbon can react with Cl or Cl_2 in the discharge, the amount of chlorine available to react with aluminum is increased. The profiles of overetched aluminum also appear susbstantially smoother with nitrogen addition. This is attributed to a different etching mechanism (ion enhanced reaction) due to substantial changes in the discharge, such as increases in ion bombardment energy and electron temperature. If the aluminum etch rate is not altered by ion bombardment energy (Smith and Bruce, 1982), the mechanism should still involve an inhibitor. Even at low CCl_4 concentrations there is substantial Cl_2 present and the sidewalls must be protected from isotropic reactions.

In this regard, it is possible that nitrogen promotes the formation of AlN on the sidewalls, and thereby inhibits lateral etching.

The use of $SiCl_4$ totally eliminates the presence of carbon introduced in the feed gas and the resulting problems (Sato and Nakamura, 1982a). It has been reported that under low pressure conditions the $SiCl_4$ discharge does not create polymer films which contaminate the sample, the chamber or the vacuum pump. This observation is contrasted by observations of residue formation with $SiCl_4$ under higher pressure etching conditions (Herb et al., 1981). In the low pressure work, lateral etching could be eliminated at high power density, suggesting that silicon deposition on the sidewalls was responsible for inhibiting the spontaneous reaction. In the presence of photoresist, no lateral etching was observed at higher pressures, indicating a participation by resist fragments in the protection chemistry. It was also found advantageous to etch at higher pressures for higher selectivities against polysilicon, silicon nitride, photoresist and silicon oxide.

In addition to control of the plasma etch chemistry and electrode potentials, reproducible aluminum etching also requires control of the aluminum film. It is well-known that oxygen is present in all deposited aluminum films, primarily as a result of residual water vapor in vacuum chambers (Faith et al., 1981). Unfortunately, since aluminum resistivity is relatively

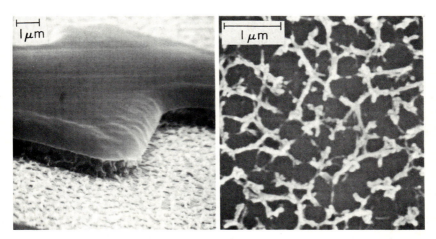

Fig. 13. SEM micrographs of the surface of an aluminum film with ~7 at% oxygen, etched at 2 Torr with a $SiCl_4$-based etch gas. The right-hand figure shows a higher magnification micrograph of the grain boundaries. (From Pender and Lindsey, 1982.)

insensitive to oxygen concentrations below 5 at% (Faith et al., 1981), this contaminant is often ignored in device processing. Concentrations in this range can, however, have a significant effect on the plasma etch properties of aluminum films. For instance, the aluminum etch rate in a $SiCl_4$ plasma has been found to decrease with increasing oxygen content of the film as measured by Auger analysis (Pender and Lindsey, 1982). This effect necessitates increased sheath potentials for high oxygen content films, if a specific etch rate is to be maintained. In turn, the increased ion energies can cause excessive resist degradation.

Oxygen in aluminum films tends to segregate at grain boundaries. Therefore, the grains etch preferentially to the grain boundaries, leaving a 'stencil' of the grains (fig. 13) as the aluminum etch process nears completion. Indeed, the higher etch rate of aluminum grains was reported in early publications on aluminum plasma etching (Schaible et al., 1978). When large concentrations of oxygen are present, complete etching of the grain boundary regions requires overetching, and in many cases, an alteration of pattern dimensions or attack of the substrate in open areas results.

3.1.3. Contaminants

The primary contaminants observed in the plasma during aluminum etching are water vapor and oxygen. These species arise principally from small leaks in the vacuum system, and/or from outgassing of wafer surfaces or hygroscopic materials (e.g., $AlCl_3$) adsorbed inside the etch chamber. Their presence in the plasma atmosphere results in uncontrollable initiation times, and even lowered aluminum etch rates in some cases (Le Claire, 1979; Tokunaga and Hess, 1980; Fok, 1980; Ranadive and Losee, 1980; Winkler et al., 1981; Donohoe, 1982; Tsukada et al., 1982). An example of the increased initiation period in CCl_4 plasmas is shown in fig. 14 as a function of H_2O content in the gas phase (Tsukada et al., 1982). Clearly no etching was possible with H_2O partial pressures above 0.5 Pa. Such effects appear to be caused by the high reactivity of aluminum surfaces to oxidants, and possibly to reactions of etch gas fragments involved in the etching of the native oxide with H_2O or O_2.

Due to the deleterious role of H_2O and O_2, these contaminants must be carefully excluded from the etch chamber, or aluminum etch gases must be chosen so that effective scavenging of such species takes place. The former approach generally requires the use of a high integrity vacuum system with a 'load-lock' to maintain etch chamber isolation from ambient.

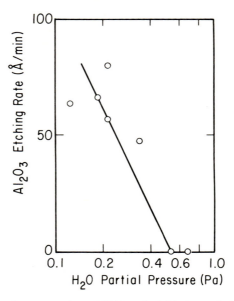

Fig. 14. Effect of small concentrations of H_2O on the initiation period (designated as Al_2O_3 etching rate). (After Tsukada et al., 1982; reprinted by permission of the publisher, The Electrochemical Society, Inc. This figure was originally presented at the Fall 1981 Meeting of the Electrochemical Society, held in Denver, Colorado).

On the other hand, scavenging is generally accomplished by the use of etch gases such as CCl_4, BCl_3, or $SiCl_4$. All of these species, or their plasma-derived fragments are capable of reacting with H_2O or O_2 to form oxides or oxychlorides, thereby removing oxygen from the plasma and inhibiting aluminum oxidation.

Differences in the efficiency of oxygen scavenging by BCl_3 and CCl_4 have been observed (Tokunaga et al., 1981). In these studies, 1% (by volume) O_2 was added to CCl_4 and to BCl_3 plasmas, and the effluent gases were monitored downstream of the reactor with a mass spectrometer. As shown in fig. 15, the O_2^+ peak decreases when a plasma is ignited both in O_2/CCl_4 and in O_2/BCl_3 mixtures. However, O_2 still remains in the O_2/CCl_4 plasma, whereas no O_2 is detected with the O_2/BCl_3 plasma. These results, indicate that BCl_3 or its plasma-generated fragments are more efficient oxygen (and apparently water vapor) scavengers than similar species in CCl_4 plasmas. Further, BCl_x species can compete effectively with aluminum surfaces for oxygen and water vapor, since small (< 4%) oxygen

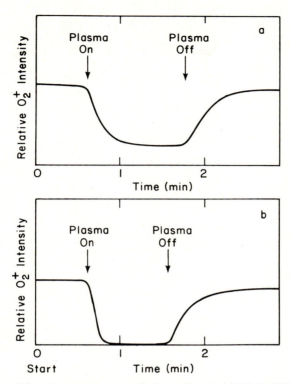

Fig. 15. Relative O_2^+ peak intensity versus discharge time in (a) 1% O_2/99% CCl_4 plasma and (b) 1% O_2/99% BCl_3 plasma. (After Tokunaga et al., 1981; reprinted by permission of the publisher, The Electrochemical Society, Inc.)

additions increase the etch rate of aluminum in BCl_3 plasmas, albeit at the expense of a slightly increased initiation time (Tokunaga et al., 1981). Indeed, O_2 has been used as an additive to a BCl_3-based etch gas for reproducible Al–Si etching (Saia and Gorowitz, 1983). Carbon tetrachloride is less efficient in oxygen scavenging; thus, the initiation period can be difficult to control with CCl_4 plasmas if H_2O or O_2 are not carefully excluded from the etch chamber.

Silicon tetrachloride, like BCl_3, readily hydrolyzes. As a resutlt, $SiCl_4$ or $SiCl_x$ should be effective scavengers for H_2O or O_2. Indeed, $SiCl_4$ displays a reproducible initiation period analogous to BCl_3 (Herb et al., 1981).

Although BCl_3 and $SiCl_4$ are capable of effective H_2O and O_2 scavenging, good vacuum integrity is still necessary for controllable aluminum etching. Oxidation of these etch gases can result in B_2O_3 (or boric acid) and SiO_2 residues which interfere with aluminum etching and may cause particulate contamination. At least in the case of $SiCl_4$, residues appear minimal (Sato and Nakamura, 1982a) and those that form can apparently be removed easily by wiping the electrodes and chamber with a diammonium phosphate solution (Herb et al., 1981).

Another residue which can form when carbon-containing vapors are used in aluminum etching involves polymerization of chlorocarbon fragments. The polymer that forms during aluminum etching in chlorocarbon vapors is not a pure chlorocarbon. Rather, this material often contains a large amount of aluminum (Nagy and Hess, 1982). Since $AlCl_3$ has a relatively low vapor pressure, it is deposited on surfaces within the reactor while etching and polymerization simultaneously occur. The result of this process is a chlorocarbon polymer containing $AlCl_3$. When an oxygen plasma is used in an attempt to strip the polymer, or the chamber is exposed to water vapor (ambient), the $AlCl_3$ is oxidized to Al_2O_3, thereby forming an inorganic/organic matrix. Such a material is difficult to remove from reactor surfaces.

3.2. Process requirements

For commercial applications a process must be optimized for productivity and quality. In evaluating productivity, several factors are involved. Not only must the number of wafers etched per hour per machine cost be considered, but the degree of automation, maintenance requirements, and floor space requirements of the machine also need to be factored into the evaluation. The quality of the process is of paramount importance. In addition to profile control, the process must also be selective against etching other materials exposed to the plasma during various stages of the process. The process must be controllable with uniform etch rates across a wafer, between wafers in a batch run, and between runs. Clearly, the process should do minimum damage, either permanent or temporary, to the etching equipment.

3.2.1. High versus low pressure etching

At present, etching can be accomplished in two operational modes: a batch mode where several wafers are etched simultaneously, or a single wafer etching (SWE) mode. Obviously, to achieve similar productivity, the processes must have a considerably higher etch rate in the SWE machines than in batch reactors. This constraint has necessitated high pressures ($\sim$ 100 Pa) in the SWE processes (Reichelderfer, 1982). These processes can use higher power densities without thermal degradation of resists due to additional thermal conduction possible at higher pressures. SWE reactors utilize a symmetric electrode configuration since at these higher pressures the self-bias generated with an asymmetric configuration is small. For batch etching processes, the etch rate can be lower and the low pressure RIE mode can be used. As has been previously discussed, the aluminum etching reaction in chlorine is spontaneous so the higher ion bombardment energy achievable in the RIE mode is not useful in increasing the aluminum etch rate if residues do not exist. The additional ion bombardment energy can be useful, however, in removing nonvolatile metals alloyed with the aluminum (i.e., copper) as will be discussed in a later section.

Recently, anisotropic etching with high ($>$ 1 μm/min) aluminum etch rates have been achieved at pressures on the order of 100 mTorr (Okano et al., 1982). These results are achieved by the use of a magnetron discharge to enhance ionization efficiency, thus increasing ion flux and etch product generation. In addition, the ion energy remains low, thereby reducing resist degradation. Finally, since passivant species are thought to be reaction products, magnetron operation should increase passivant formation without increasing resist erosion.

The quality of the etch is determined by the control and reproducibility of the process. The factors to be considered include etch uniformity, etch selectivity of aluminum to other materials, and feature profiles. At the present time, no definitive comparison of batch and SWE processes has been reported for aluminum etching. This may be due to the limited amount of plasma aluminum etching in production, particularly using SWE. It is possible, however, to discuss process directions which would be expected to provide better selectivity over photoresist, silicon, and silicon dioxide.

3.2.2. *Selectivity*

A successful etch process requires that masking materials used for pattern transfer withstand the reactive plasma atmosphere without significant erosion or degradation; otherwise, pattern sizes will be altered. In addition, attack of the underlying film or substrate by the etchant after the etch endpoint is reached must be minimal. Naturally, these stipulations require careful attention to the etch chemistry.

Silicon with low dopant concentration will etch in a Cl_2 discharge, but the reaction is ion enhanced and thus the reaction rate is proportional to the ion bombardment energy. Doped polysilicon, however, will etch in both an ion enhanced and a spontaneous (purely chemical) mode. The spontaneous etch rate can be reduced only by reducing the etchant concentration. Unfortunately, this also reduces the aluminum etch rate and hence does not affect the selectivity. Enhanced silicon etching can be reduced by etching under conditions of low ion bombardment energy. Processes which are anisotropic under low powers at high pressures, therefore, should have better selectivities to silicon etching. The additives used in such processes can have a unique power threshold below which isotropic etching occurs. For instance, the threshold power for $CHCl_3$ is higher than for CH_3Cl when used as an additive with Cl_2 at high pressures and power densities (Smith and Saviano, 1982). This suggests that inhibitor formation with CH_3Cl requires less molecular dissociation than with $CHCl_3$. A study of the plasma chemistry of these systems (Smith and Saviano, 1982) indicates that hydrogen abstraction by chloride atoms is a dominant reaction. That is, reaction of chlorine atoms with $CHCl_3$ and with CH_3Cl seems to occur in the following way:

$$Cl + CHCl_3 \rightarrow HCl + CCl_3, \tag{10}$$

$$3Cl + CH_3Cl \rightarrow 3 HCl + CCl. \tag{11}$$

The detection of these reaction products (Smith and Saviano, 1982) implies that HCl or polymer formed from CCl species contributes to the sidewall inhibition.

Silicon dioxide has a negligible etch rate in a Cl_2 discharge under most conditions. In the presence of additives used in mixtures to improve etch quality and initiation times, the SiO_2 etch rate can be increased. Since this is also an enhanced process, the rate increase is again proportional to ion bombardment energy.

Momentum transfer due to ion bombardment causes temperature increases and bond breaking in resist materials used as masking layers. In addition, reactive chlorine species can abstract hydrogen from resists. The result of these processes is severe degradation and flow of the resist, thereby altering pattern dimensions in the etched film. Such effects are particularly serious in aluminum etching with CCl_4, where the etch rate of positive resists (generally novolac resins) processed according to manufacturers' specifications, is essentially equal to that of the aluminum film. Interestingly, polysilicon films can be etched in CCl_4 plasmas without significant resist degradation (Robb, 1979). Such disparate observations in aluminum and polysilicon etching are probably due to differences in the etch products between these two materials (Chapman and Nowak, 1980).

Aluminum trichloride is a Lewis acid, and is therefore highly reactive toward organic materials. Further, the relatively low volatility of $AlCl_3$ promotes deposition of this etch product onto resist surfaces. Indeed, Auger depth profiles of positive resist residues after aluminum etching and plasma ashing show the presence of aluminum throughout the carbonaceous film (Tracy and Mattox, 1982). Most likely, $AlCl_3$ incorporated into the resist attacks the phenol group in the novolac resin, ultimately resulting in carbon–carbon bond breakage in the polymer chain; thus resist degradation occurs (Hess, 1982).

In addition to degradation and erosion, resist swelling has been reported during aluminum etching. Because of the direct relationship between apparent pattern undercut and resist swelling, and the effects of argon addition on swelling and on wafer temperature, this effect has been attributed to the heat of reaction between neutral etchant species and aluminum (Horiike et al., 1982). In these studies, resist swelling could be minimized by properly cooling the wafers.

Prevention of resist degradation and distortion is mandatory if polymer materials are to be used for pattern definition in aluminum plasma etching. As a result, considerable effort has gone into the development of processes which either reduce the surface temperature of resists, or which 'harden' or 'toughen' positive (usually AZ-type) resists.

Temperature increases can be minimized by providing improved gas phase thermal conductivity or by enhancing contact between the wafer and the cooling medium. In the former approach, higher pressures are advantageous. The addition of inert gases (especially helium) to the etchant mixture can improve thermal conduction, and thus promote cooling. Contact improvement can be accomplished in several ways, the most promising of which utilizes direct contact between the cooling fluid and the backside of the wafer (Egerton et al., 1982).

Resist attack is often diminished if pre-treatments are invoked to 'toughen' the resist against plasma exposure. These treatments utilize either a UV exposure of the resist or a CF_4 plasma treatment, in both cases followed by a high temperature ($> 140°C$) bake cycle (Donohoe, 1982;

Tracy and Mattox, 1982). It should be noted that long (15–300 min, depending upon the temperature) bake cycles alone can also harden resists, but are often unsuitable because of the process time required (Donohoe, 1982). In general, bake cycles are presumed to drive off residual cellosolve acetate and to decompose the diazide sensitizer present in positive resists (Donohoe, 1982).

Deep UV exposure (~ 250 nm) of positive resists has been shown to be effective in reducing resist flow during subsequent thermal treatments. Apparently, in the presence of water, the UV causes surface crosslinking of the resist and/or decomposes the diazide sensitizer, thereby forming a hardened layer which prevents resist flow and thus stabilizes the pattern (Hiraoka and Pacansky, 1981; Allen et al., 1982). Stabilization is necessary so that elevated temperature ($> 140°C$) bake cycles can be used prior to etching. In addition, reaction and penetration of the resist by $AlCl_3$ is probably reduced after such treatments.

Nitrogen or CF_4 plasma exposure of resist surfaces also hardens the material against aluminum etchants. With N_2, ion bombardment and UV from the plasma appear to form a surface 'skin' similar to that proposed in UV flood exposure (Moran and Taylor, 1981). Plasmas of fluorine-containing gases (usually CF_4) apparently react with positive resist materials to form a fluorinated and probably crosslinked surface layer which prevents flow and minimizes reactivity (Ma, 1980; Dobkin and Cantos, 1981).

The above treatments of positive resist prior to aluminum plasma etching in CCl_4-based gases permit selectivities between 4:1 and 10:1 (aluminum:resist) to be achieved (Tracy and Mattox, 1982; Donohoe, 1982). However, the selectivity observed is extremely sensitive to resist processing and to etching conditions. In particular, the use of $SiCl_4$ as an inhibitor in an etch gas mixture is reported to give similar selectivities (as well as anisotropic profiles) without the need for resist pre-treatments (Smith and Yao, 1983). The reason for such results may be related to the selective deposition of silicon onto the resist surface at rates too high for ion removal. The deposited layer then protects the resist surface from chemical attack.

Some investigations have indicated that polymeric resists may play an important role in generating anisotropic profiles during aluminum etching (Oda and Hirata, 1980; Sato and Nakamura, 1982a). That is, sputtering or degradation of polymer resists can result in sidewall coatings which inhibit lateral etching because of the lack of ion bombardment in this region. However, use of etch mixtures containing inhibitor species has eliminated the need for such anisotropy mechanisms. In fact, some interest has developed in the use of inorganic materials as masking layers so that degradation of the resist can be avoided.

4. Aluminum alloy etching

The metallization presently used in VLSI is not pure aluminum but alloys of aluminum with silicon and/or copper. The alloying is necessary to minimize metallization problems such as spiking at contact interfaces due to the solubility of silicon in aluminum, electromigration at high current densities, and hillock formation (Learn, 1976). Since silicon is volatilized readily by ion bombardment in a Cl_2 discharge, the aluminum–silicon alloy is generally etched without substantial problems.

Etching of the copper alloy is much more difficult because $CuCl_2$ and $CuCl$ are not volatile (Broydo, 1983). Elevated temperatures (> 40°C) and/or high energy ion bombardment are necessary to remove the copper. Alternatively, post-etching treatments which utilize concentrated nitric acid to dissolve copper chlorides can be used (Herndon and Burke, 1977; Nakamura et al., 1981).

In high pressure etching, 2% Cu–Al alloy has been etched with a Cl_2–inhibitor mixture at temperatures in excess of 200°C using SiO_2 masking. In these circumstances it has been observed that residue can be eliminated if $SiCl_4$ is used as an inhibiting additive while substantial residue remains if a carbon containing additive is used. In these studies, it has been postulated that copper catalyzes the deposition of carbon to form a residue (Smith and Yao, 1983).

5. Corrosion

Exposure of plasma-etched aluminum films to ambient conditions generally results in corrosion of the aluminum. The cause of this phenomenon is hygroscopic chlorine-containing residues (primarily $AlCl_3$) remaining on the aluminum sidewalls, on the substrate surface, or in the resist. In the presence of moisture, these residues hydrolyze to generate HCl, which corrodes aluminum. Removal of chlorinated species is imperative for effective passivation, because it appears that no long-term passivation is possible while chlorine is in contact with the aluminum film. Presumably, a catalytic effect is operative wherein chlorine species continue to promote corrosion even when the source of chlorine has been removed. Further, corrosion is reported to be further enhanced by carbon contamination and by radiation damage resulting from plasma exposure (Lee et al., 1981). Thus, in order to eliminate corrosion, chlorinated compounds must be removed from film surfaces, and the passivating native oxide, which was removed during etching, re-grown on the aluminum surface.

Since merely removing the resist after etching is not adequate, a number of post-etching treatments have been used to inhibit aluminum corrosion. Oxygen plasma cycles and de-ionized water rinses greatly reduce chlorine concentrations; unfortunately, some chlorine still remains on surfaces (Kitcher, 1980). Also, removal of residual chlorine by hydrogen plasma treatments has been reported to reduce corrosion (Ranadive and Losee, 1981). Elimination of post-etch corrosion has been achieved by oxidation of aluminum surfaces in dry oxygen at temperatures between 300 and 350°C (Lee et al., 1981). It is believed that such treatments remove carbon and chlorine contamination, anneal radiation damage, and re-grow the protective oxide layer. Another method of achieving the same basic result is to expose the etched wafers to hot ($> 200°C$) air followed by de-ionized water rinsing (Horiike et al., 1982).

Another post-etch treatment which removes chlorine-containing compounds and supplies surface passivation involves the use of a fluorocarbon plasma (Fok, 1980; Donelly and Flamm, 1981; Herndon et al., 1982). Here, the chloride residues are converted into nonhygroscopic fluorides by reaction with fragments derived from CF_4, SF_6, etc. This procedure eliminates chlorine, and permits exposure of the etched aluminum to atmospheric conditions. Sulfur hexafluoride seems to be especially suitable for such treatments, apparently due to the avoidance of carbon contamination (Meisner, 1982). In some cases, the fluoride layer is subsequently removed and a passivating oxide layer grown with a fuming nitric acid rinse (Fok, 1980).

Copper additions to aluminum films appear to cause accelerated post-etch corrosion. At least two factors are probably involved with this process. Due to its low vapor pressure, hygroscopic $CuCl_2$ is probably left in contact with aluminum films. In addition, much of the copper is present in the aluminum grain boundaries as $CuAl_2$. Thus, the grain boundaries assume a cathodic potential relative to the aluminum grains (Van Horn, 1967; Lee et al., 1981). Atmospheric water adsorption and subsequent hydrolysis of $CuCl_2$ and $AlCl_3$ to generate HCl and chloride ions therefore result in electrolytic corrosion.

6. Safety aspects

Many of the gases and vapors used in aluminum plasma etching are toxic, carcinogenic, and/or corrosive. For instance, CCl_4 and $CHCl_3$ are toxic and suspected carcinogens, while Cl_2 and BCl_3 are highly corrosive and irritating to mucous membranes. Both BCl_3 and $SiCl_4$ hydrolyze readily, generating HCl. In addition, because BCl_3 is often prepared by the reduction of boric oxide with carbon followed by chlorination, or by the chlorination of boron carbide, phosgene ($COCl_2$) is generally present at concentrations between 25 and 800 ppm in BCl_3 source bottles (Herb, 1982). Clearly, safe use of such materials requires careful attention to the design and operation of gas handling systems.

Unfortunately, adherence to proper safety precautions for inlet vapors does not ensure overall process safety. Due to the complex decomposition reactions occurring in r.f. glow discharges, a number of hazardous materials can be synthesized during the etching process. The major gases and vapors present in the effluent from CCl_4 discharges are shown in table 1 (section 2.1). Of these materials, carbon tetrachloride and hexachloroethane (C_2Cl_6) are suspected carcinogens. In addition, minor amounts of trichloroethylene (C_2HCl_3) have been detected in CCl_4 plasma effluents (Herb, 1982). Also if water vapor and/or oxygen are present in the plasma atmosphere, $COCl_2$ and HCl are generated (Hess, 1981). Further, analysis of solid residues from the etch chamber indicates the presence of C_2Cl_6, C_4Cl_6, C_4Cl_8, and still higher molecular weight chlorocarbons, along with aluminum, chloride, and iron inorganic compounds (Herb, 1982). Some of these materials can be released or can hydrolyze to form HCl when the etch chamber is exposed to the atmosphere. As a result, extensive system purging and/or post-etch treatments such as hydrogen or fluorine discharges should be carried out prior to opening the chamber to avoid operator exposure to hazardous vapors.

Because of the dangerous materials used or generated during aluminum etching, extreme caution must be exercised when cleaning the reactor chamber or reactor fixtures, and particularly when changing pump oil. Carcinogenic chlorocarbon vapors and HCl are dissolved in most oils; further, these compounds are often concentrated by pump oils. Therefore, during routine pump (and cold trap) maintenance, and oil changes, oil or residue contact with the skin, and inhalation of vapors from cold traps, oil filters, and pumps, must be avoided.

Pump oils are often degraded or polymerized by reactive compounds in the reactor effluent. Lewis acids such as $AlCl_3$ and BCl_3 can react with silicone pump oils, causing degradation (Lehman et al., 1981). With hydrocarbon-based oils, reactive chlorine species can remove a hydrogen atom from a hydrocarbon molecule, resulting in a free radical which initiates polymerization (O'Hanlon, 1981). To avoid decomposition and polymerization, perfluoropolyethers are often used as pump oils (O'Hanlon, 1980). These fluids show little reactivity, even with Lewis acids, unless elevated ($> 100°C$) temperatures are invoked.

Reactive source and product species cause considerable difficulties for pumping systems. For instance, oxidation of $AlCl_3$, BCl_3, or $SiCl_4$ results in particulate (Al_2O_3, B_2O_3, and SiO_2) formation. These particles, along with polymeric residues in the oil, can cause bearing failure by plugging lubrication ducts (O'Hanlon, 1981). It is thus generally advisable to incorporate an oil filter to prolong pump life, and to extend the interval between oil changes.

7. Acknowledgments

The authors are grateful to S. Broydo, D.A. Danner, A. Keaton, E. Sirkin, D.L. Smith, and C.C. Tang for helpful comments and for critical readings of the manuscript. D.W. Hess acknowledges support of his aluminum etching work by the National Science Foundation under Grant No. ECS-8021508.

References

Allen, R., M. Foster and Y.T. Yen, 1982, J. Electrochem. Soc. **129**, 1379.
Broydo, S., 1983, Solid State Technol. **26** (4), 159.
Bruce, R.H., 1981a, Solid State Technol. **24** (10), 64.
Bruce, R.H., 1981b, J. Appl. Phys. **52**, 7064.
Bruce, R.H., 1981c, in: Plasma Processing, eds. R.G. Frieser and C.J. Mogab (The Electrochemical Society, Pennington), p. 243.
Bruce, R.H., and G. Malafsky, 1982, in: Plasma Processing, eds. J. Dielman, R.G. Frieser and G.S. Mathad (The Electrochemical Society, Pennington) p. 336.
Bruce, R.H., G.P. Malafsky, A.R. Reinberg and W.W. Yao, 1981, in: Technical Digest of the International Electron Devices Meeting, Washington, D.C. (IEEE, New York) p. 578.
Buchel'nikova, I.S., 1959, Sov. Phys. JETP **35**, 783.
Chapman, B.N., and M. Nowak, 1980, Semicond. Int. **3** (10), 139.
Christophorou, L.C., and J.A. Stockdale, 1968, J. Chem. Phys. **48**, 1956.
Coburn, J.W., and E. Kay, 1972, J. Appl. Phys. **43**, 4965.
Coburn, J.W., and H.F. Winters, 1979, J. Appl. Phys. **50**, 3189.

Curtis, B.J., 1980, Solid State Technol., **23** (4), 129.
Dobkin, D.M., and B.D. Cantos, 1981, Electron Device Lett. **EDL-2**, 222.
Donelly, V.M., and D.L. Flamm, 1981, Solid State Technol. **24** (4), 161.
Donohoe, K.G., 1982, in: Plasma Processing, eds. J. Dielman, R.G. Frieser and G.S. Mathad (The Electrochemical Society, Pennington) p. 306.
Egerton, E.J., A. Nef, W. Millikin, W. Cook and D. Baril, 1982, Solid State Technol. **25** (8), 84.
Faith, T.J., R.S. Irven, J.J. O'Neill Jr. and F.J. Tams, III, 1981, J. Vac. Sci. Technol. **19**, 709.
Flamm, D.L., V.M. Donelly and D.E. Ibbotson, 1983, J. Vac. Sci. Technol., to be published.
Fok, Y.T., May 1980, Electrochem. Soc. Extended Abstracts, St. Louis, MO Meeting, Abstract No. 115.
Fraser, D.B., 1978, in: Thin Film Processes, eds. J.L. Vossen and W. Kern (Academic, New York) p. 115.
Hawkins, D.T., 1979, J. Vac. Sci. Technol. **16**, 1051.
Heiman, N., V. Minkiewicz and B. Chapman, 1980, J. Vac. Sci. Technol. **17**, 731.
Herb, G.K., 1982, Tegal Plasma Symposium (Semicon-East Technical Program).
Herb, G.K., R.A. Porter, P. Cruzan, D. Agraz-Guerena and B.R. Soller, October 1981, Electrochem. Soc. Extended Abstracts, Denver, CO Meeting, Abstract No. 291.
Herndon, T.O., and R.L. Burke, 1977, Kodak '77 Interface Proceedings.
Herndon, T.O., R.L. Burke and J.F. Howard, 1982, Kodak '82 Interface Proceedings.
Hess, D.W., 1981, Solid State Technol. **24** (4), 189.
Hess, D.W., 1982, Plasma Chem. Plasma Process. **2**, 141.
Hiraoka, H., and J. Pacansky, 1981, J. Electrochem. Soc. **128**, 2645.
Horiike, Y., T. Yamazaki, M. Shibagaki and T. Kurisaki, 1982, Japan. J. Appl. Phys. **21**, 1412.
Keaton, A., and D.W. Hess, 1983, unpublished results.
Kitcher, J.R., October 1980, Electrochem. Soc. Extended Abstracts, Hollywood, FL Meeting, Abstract No. 329.
Learn, A.J., 1976, J. Electrochem. Soc. **123**, 894.
Le Claire, R., 1979, Solid State Technol. **22** (4), 139.
Lee, R.E., 1979, J. Vac. Sci. Technol. **16**, 164.
Lee, W.Y., J.M. Eldridge and G.C. Schwartz, 1981, J. Appl. Phys. **52**, 2994.
Lehmann, H.W., E. Herb and K. Frick, 1981, Solid State Technol. **24** (10), 69.
Ma, W.H.L., 1980, IEDM Proceedings (Washington, D.C.) p. 574.
Marcoux, P.J., and P.D. Foo, 1981, Solid State Technol. **24** (4), 118.
Meisner, M., 1982, Synertek Corporation, private communication.
Minford, J.D., M.H. Brown and R.H. Brown, 1959, J. Electrochem. Soc. **106**, 185.
Mogab, C.J., 1980, in: Physics of Dielectric Solids, 1980, ed. C.H.L. Goodman (The Institute of Physics, Bristol) p. 37.
Moran, J.M., and G.N. Taylor, 1981, J. Vac. Sci. Technol. **19**, 1127.
Mothes, K.G., E. Schultes and R.N. Schindler, 1972, J. Phys. Chem. **76**, 3758.
Mucha, J.A., and D.W. Hess, 1983, in: Introduction to Microlithography, eds. L.F. Thompson, C.G. Willson and M.J. Bowden (ACS Symposium Series, No. 219, American Chemical Society, Washington, D.C.).
Nagy, A.G., and D.W. Hess, 1982, J. Electrochem. Soc. **129**, 2530.
Nakamura, M., M. Itoga and Y. Ban, 1981, in: Plasma Etching, eds. R.G. Frieser and C.J. Mogab (The Electrochemical Society, Pennington) p. 225.

Nordine, P.C., and J.D. Le Grange, 1976, AIAA Journal **14**, 644.
Oda, M., and K. Hirata, 1980, Japan. J. Appl. Phys. **19**, L405.
O'Hanlon, J.F., 1980, A User's Guide to Vacuum Technology (Wiley–Interscience, New York) p. 163.
O'Hanlon, J.F., 1981, Solid State Technol. **24** (10), 86.
Okano H., T. Yamazaki and Y. Horiike, 1982, Solid State Technol. **25** (4), 166.
Pender, M., and P.C. Lindsey, 1982, Zylin Corporation, unpublished results.
Poulsen, R.G., H. Nentwich and S. Ingrey, 1976, IEDM Proceedings (Washington D.C.) p. 205.
Ranadive, D.K., and D.L. Losee, 1980, in: Plasma Processing, eds. R.G. Frieser and C.J. Mogab (The Electrochemical Society, Pennington) p. 236.
Reichelderfer, R.F., 1982, Solid State Technol. **25** (4), 160.
Robb, F., 1979, Semicond. Int. **2** (12), 60.
Saia, R.J., and B. Gorowitz, 1983, Solid State Technol. **26** (4), 247.
Sato, M., and H. Nakamura, 1982a, J. Vac. Sci. Technol. **20**, 186.
Sato, M., and H. Nakamura, 1982b, J. Electrochem. Soc. **129**, 2522.
Schaible, P.M., W.C. Metzger and J.P. Anderson, 1978, J. Vac. Sci. Technol. **15**, 334.
Sinha, A.K., in: VLSI Science and Technology, 1982, eds. C.J. Dell'Oca and W.M. Bullis (The Electrochemical Society, Pennington) p. 173.
Smith, D.L., and R.H. Bruce, 1982, J. Electrochem. Soc. **129**, 2045.
Smith, D.L., and P.G. Saviano, 1982, J. Vac. Sci. Technol. **21**, 768.
Smith, D.L., and W. Yao, 1983, private communication.
Smith, T., 1972, Surf. Sci. **32**, 527.
Spencer, J.E., and B.Y. Shu, 1982, in: VLSI Science and Technology, 1982, eds. C.J. Dell'Oca and W.M. Bullis (The Electrochemical Society, Pennington) p. 103.
Stull, D.R., 1947, Ind. Eng. Chem. **39**, 517.
Thornton, J.A., and A.S. Penfold, 1978, in: Thin Film Processes, eds. J.L. Vossen and W. Kern (Academic, New York) p. 76.
Tiller, H.J., D. Berg and R. Mohr, 1981, Plasma Chem. Plasma Process. **1**, 247.
Tokunaga, K., and D.W. Hess, 1980, J. Electrochem. Soc. **127**, 928.
Tokunaga, K., F.C. Redeker, D.A. Danner and D.W. Hess, 1981, J. Electrochem. Soc. **128**, 851.
Tracy, C.J., and R. Mattox, 1982, Solid State Technol. **25** (6), 83.
Tsukada, T., H. Mito and M. Nagasaka, 1982, in: Plasma Processing, eds. J. Dielman, R.G. Frieser and G.S. Mathad (The Electrochemical Society, Pennington) p. 326.
Van Horn, K.R., ed., 1967, Aluminum, Vol. I (American Society for Metals, Metals Park, OH), p. 209.
Vedeneyev, V.J., L.V. Gurvich, V.N. Kondrat'yev, V.A. Medvedev and Ye.L. Frankevich, 1966, Bond Energies, Ionization Potentials, and Electron Affinities (St. Martin's Press, Dunmore).
Vossen, J.L., and J.J. Cuomo, 1978, in: Thin Film Processes, eds. J.L. Vossen and W. Kern (Academic, New York) p. 11.
Waits, R.K., 1978, in: Thin Film Processes, eds. J.L. Vossen and W. Kern (Academic, New York) p. 131.
Winkler, U., F. Schmidt and N. Hoffman, 1981, in: Plasma Processing, eds. R.G. Frieser and C.J. Mogab (The Electrochemical Society, Pennington) p. 253.
Wortik, T., R. Moore and H.J. Schlesinger, 1949, J. Am. Chem. Soc. **71**, 3265.

CHAPTER 2

PLASMA ETCHING OF REFRACTORY GATES OF METALS, SILICIDES AND NITRIDES

T. PAUL CHOW

General Electric Company
Corporate Research and Development
Schenectady, NY 12345, USA

A.N. SAXENA

American Microsystems, Inc.
A Subsidiary of Gould Inc.
Santa Clara, CA 95051, USA

L.M. EPHRATH and R.S. BENNETT

IBM Corporation
Hopewell Junction, NY 12533, USA

Dry Etching for Microelectronics, edited by R.A. Powell
© *Elsevier Science Publishers B.V., 1984*

Contents

1. Introduction — 41
   1.1. Dominance of interconnects — 41
   1.2. Refractory metals and metal silicides — 43
   1.3. General etching considerations — 45
2. Pattern definition of single-level gates — 50
   2.1 Metals — 50
      2.1.1. Fluorinated plasmas — 51
      2.1.2 Chlorinated plasmas — 56
      2.1.3. Halide mixtures — 58
   2.2. Metal silicides — 58
      2.2.1. Fluorinated plasmas — 59
      2.2.2. Chlorinated plasmas — 60
      2.2.3. Fluorinated/chlorinated mixtures — 62
   2.3. Metal nitrides — 63
3. Pattern definition of composite gates — 63
   3.1. Polycide — 63
      3.1.1. Fluorinated plasmas — 64
      3.1.2. Chlorinated plasmas — 65
      3.1.3. Halide mixtures — 66
   3.2. Metal-nitride/metal gates — 67
4. Etching dependence on thin film properties — 67
5. Endpoint detection — 68
6. Mechanisms — 70
7. Summary — 73
References — 75

1. Introduction

Refractory metals were offered as an alternative to doped polycrystalline silicon (poly-Si) in the early days of self-aligned MOS gate technology. However, at that time, the process simplicity and flexibility offered by poly-Si more than offset the higher conductivity of these metals. Recently, due to the increasing interconnection line lengths and decreasing line widths in high-speed very-large-scale-integrated circuits (VLSIC), the conductivity of doped poly-Si is becoming a limiting factor in determining the overall circuit performance. Consequently, the interests in microelectronic applications of refractory metals have been revived. The silicides of these metals, which possess some of the desirable properties of both the metals and poly-Si, have also attracted much attention. Because the process compatibility with the current production technologies is easier with the silicides than with the metals, the silicides are being implemented rapidly.

1.1. Dominance of interconnects

The length and the resistivity of the interconnect layer are the key factors which will dominate the overall performance of the future VLSIC's. For a detailed discussion of this subject, see Saraswat and Mohammadi (1982), and Sinha et al. (1982). However, an elementary model, to illustrate the effect of the length, l, and the sheet resistivity, $\varrho_\square$, of the interconnect layer on the RC time delay is given below (Saxena and Pramanik, 1983). In fig. 1, the resistance, R, of a conductor, whose length is l, width is w, and thickness is d, is given by

$$R = \varrho \, \frac{l}{dw} \, \Omega, \tag{1}$$

where ϱ is the thin-film resistivity of the conductor material deposited in the form of a film. If this conductor is sitting on top of an oxide or dielectric layer of thickness t_{ox} and dielectric constant k, the RC time constant associated with it will be

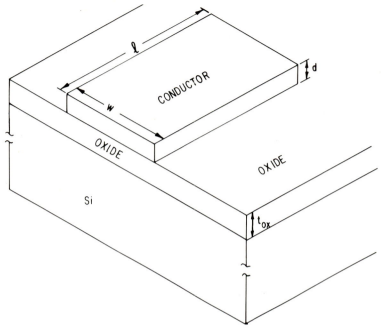

Fig. 1. Schematic of a conductor with length l, width w, and thickness d, on oxide with thickness t_{ox}.

$$RC = \varrho_\square \frac{l}{dw} kl \frac{w}{t_{ox}} = \varrho_\square l^2 \frac{k}{t_{ox}}. \qquad (2)$$

The term $\varrho_\square$, referred to as the *sheet resistivity* of the film, and related to the thin-film resistivity, ϱ, of the film as

$$\varrho_\square = \frac{\varrho}{d}, \qquad (3)$$

is commonly used in the VLSI technology while discussing the properties of the film. As it is obvious from eq. (2), the dominant factors of the interconnect layer contributing to the RC time delay are its sheet resistivity and its length l, in particular the latter because it is a squared term. As VLSIC's grow in size, the length of the interconnects increases and plays the most important role in determining the speed of the VLSIC's. For this reason, multilevel metallization becomes a requirement.

1.2. Refractory metals and metal silicides

The type of metallizations/interconnects needed in VLSIC's can be categorized as high temperature ($T > 550°C$) and low temperature ($T < 550°C$) metallizations.

Examples and uses of high temperature metallization are given in table 1 and those of low temperature metallization are given in table 2. The most prevalent use of high temperature metallization is for MOS gates, capacitors, and interconnects, for which doped polycrystalline silicon

Table 1
High temperature metallizations.

Applications/use	Present	New options
1. MOS gates and capacitors and interconnects	Doped poly-Si	1. Polycide (defined to be a two-layer structure of silicide over poly-Si) 2. Silicide 3. Refractory metals
2. N^+ and P^+ layers within bulk silicon	Implanted or diffused bulk silicon	Strapping of N^+ and P^+ layers with silicides

Table 2
Low temperature metallizations.

Applications/use	Present	New options
1. First layer metal	Al and its alloys	1. Refractory metals 2. Al and its alloys with silicide or refractory metals
2. Second and subsequent layer metal	Al and its alloys	1. Refractory metals 2. Al and its alloys with silicide or refractory metals
3. Ohmic contacts to N^+ and P^+ and/or Schottky contacts to N and P silicon	Al and its alloys, PtSi Pd_2Si with barrier metals	1. Various silicides with and without barrier metals 2. Selective W deposition
4[a]. Polysilicon gates and interconnects	Doped poly-Si	Doped poly-Si strapped with selective W deposition

[a]Poly-Si is a high-temperature metallization, however, on strapping it with W, it falls into the category of low-temperature metallization.

Table 3
Properties of high and low temperature materials.

Type of material	Material	Thin-film resistivity ($\mu\Omega$ cm)	Melting point (°C)	Chemical compatibility with VLSI processes	Thermal oxidation for passivation	LPCVD process for film deposition	Dry etching process for the films
HTM	Doped poly-Si	500-525	1410	Yes	Yes	Yes	Good
	$TiSi_2$	13- 16	1540	Yes, except etches in HF	Yes	?	Good
	WSi_2	30- 70	2165	Yes	Yes	Yes	Good
	$TaSi_2$	35- 55	2170	Yes, except etches slowly in HF	Yes	Yes	Good
	$MoSi_2$	40-100	2050	Yes	Yes	Yes	Good
	W	5- 15	3410	?	No	Yes	Good
	Mo	6- 15	2610	?	No	Yes	Good
LTM	Al	2- 5	660	?	No	?	Fair
	W	5- 15	3410	?	No	Yes	Good
	PtSi	28- 35	1229	Yes/?	Yes/?	No	None

(poly-Si) is used at present. Table 3 lists several properties of these materials. Most desirable high temperature materials which will find large usage in VLSI technologies are $TiSi_2$, WSi_2, $TaSi_2$, $MoSi_2$, W and Mo. A few other materials, e.g. $CoSi_2$ and Ti, have also been suggested for VLSI applications. However, for a variety of reasons, they have been found to be less suitable than those given above (and in table 3), except for the strapping and contact applications of Ti. As of this writing, no clear-cut advantage of one silicide over the other, even within the group of four silicides, $TiSi_2$, WSi_2, $TaSi_2$ and $MoSi_2$, has been demonstrated; however, $TiSi_2$ offers the lowest resistivity.

It is clear from the discussion above that refractory metals and metal silicides are needed primarily because their sheet resistivities are lower than that of doped poly-Si which has been used extensively up to now in integrated circuits. Furthermore, silicides are preferred over the refractory metals for several process applications because of the good oxidation properties of the former.

Polycide, a bilayer structure with silicide over poly-Si, is more difficult to pattern but offers some advantages over refractory metals or metal silicides. Polycide preserves the reliability of the poly-Si/SiO$_2$ interface while providing a direct replacement of the single-layer poly-Si gates and interconnects. A more subtle advantage has been reported by Mochizuki et al. (1980): contact resistance of buried contacts was found to increase with increasing temperatures for silicide contacts but not for polycide contacts.

1.3. General etching considerations

Etching for pattern definition is an important step in the integrated-circuit fabrication sequence. Basically, it involves the selective removal of the exposed portions of a thin film layer over the underlying layers or substrate. Conventionally, wet chemical solutions have been extensively used. While refractory metals can be readily dissolved in many oxidizing solutions, like HNO_3 and H_2O_2, the metal silicides are relatively inert to most chemicals used in device processing. The etching requirements necessary for the fabrication of VLSI circuits are particularly stringent because of the small dimensional features needed. Indeed, the wet chemical techniques have become inadequate because of their isotropic nature. Also, wetting of small openings is difficult due to capillary effects. Various dry etching methods based on reactive gas plasmas have been actively explored because of the directional characteristics of these processes.

In table 4, a summary of refractory metals on which plasma etching was reported in the literature is presented, along with the etching gases used for each. Both fluorinated and chlorinated gases as well as their mixtures have been studied. A similar table for silicides is shown in table 5. As is well known in silicon etching, one of the fundamental differences is the high volatility of the reaction products (i.e., silicon fluorides) in fluorine-based plasmas. To facilitate the desorption of the silicon chlorides so that further etching can proceed, ion-assisted reactions are necessary in chlorine-based plasmas. However, this enhances the etching anisotropy, since the sidewall of the etched film or substrate is not subjected to energetic ion bombardment. Also, oxygen is often added, particularly for the halocarbons (like CF_4 and CCl_4), to enhance the etch rate of many of these gases. It's effect is believed to be the more efficient generation of fluorine atoms or radicals (Mogab et al., 1978). Also indicated in the two tables are the reactor configurations which have been used. As the plasma technology evolved, cylindrical (barrel-type) reactors have been replaced by planar (parallel-plate) reactors. This latter group can be categorized into planar plasma etching (PPE), reactive ion etching or reactive sputter etching (RIE or RSE), and, most recently, flexible diode (FDE) and triode etching (TE) modes (Chapman, 1980; Ephrath, 1982; Vossen, 1979). Generally, PPE systems are operated at higher pressures (~100 mTorr to 1–

Table 4
Summary* of etching gases and modes used for refractory metals.

	Mo	W	Ti	Ta	Nb	V	Ti:W
CF_4	a [1, 2]		a [19]	a [22]	a [19]		
		c [13]	c [17]		c [19]		
CF_4/O_2	a [3–6]	a [5]	a [5]	a [5, 23]	a [4]		
	b [7]						
CHF_3		c [17]					
C_2F_6			c [20]			c [20, 24]	
NF_3	b [8–10]						b [10]
NF_3/Ar	b [8, 9]						
NF_3/He	b [9]						
SF_6	c [11]						
SF_6/O_2		c [18]					
CCl_2F_2	c [12]						
$C_2Cl_3F_3$	c [12]						
CCl_2F_2/O_2	c (?)[13]						
$C_2Cl_2F_4/O_2$			a [3]				
CCl_4/O_2	b [14]						
	c [15, 16]						
CF_3Br	c [23]						
$CF_3Br/O_2/He$			b [21]				

* Explanation of letters and numbers used:
(a) Barrel or downstream plasma etching;
(b) Planar plasma etching;
(c) Reactive ion etching, reactive sputter etching;
(?) Reference ambiguous on reactor configuration and/or specific gas used.
References:
[1] Maeda and Fujino (1975); [2] Takahashi et al. (1978); [3] Legat and Schilling (1975); [4] Horiike and Shibagaki (1977); [5] Poulsen (1977); [6] Chow (1982); [7] Shah (1979); [8] Chow and Steckl (1980b); [9] Chow and Steckl (1982); [10] Roth and Hagen (1980); [11] Bensaoula et al. (1983); [12] Hosokawa et al. (1974); [13] Fukumoto et al. (1981); [14] Hirata et al. (1981); [15] Kurogi and Kamimura (1982); [16] Gorowitz and Saia (1982); [17] Lehmann and Widmer (1979); [18] Randall and Wolfe (1981); [19] Harada et al. (1981); [20] Matsuo (1978); [21] Mogab and Shanoff (1977); [22] Busta et al. (1979); [23] Dybwad (1983); [24] Matsuo (1980b).

5 Torr) than RIE, FDE and TE systems (10–100 mTorr. For comparison, conventional sputter etching uses inert gases at 1 to ~10 mTorr. As can be seen in fig. 2, the main configuration difference between the PPE and

Table 5
Summary* of etching gases and modes used for refractory metal silicides.

	$MoSi_2$	WSi_2	$TaSi_2$	$TiSi_2$	$NbSi_2$
CF_4/O_2	a [1–3]		a (?)[17]		a [22]
		b [12]	b [18]		
	c [4](?)	c [13, 14]			
CF_4/C_2F_6	c [3]				
NF_3	b [5–7]				
NF_3/Ar	b [6]				
NF_3/He	b [6]				
SF_6/O_2		b [15]			
	c [8, 9]				
CCl_4/O_2				b (?)[21]	
	c [10]				
BCl_3/Cl_2			[19]		
CCl_2F_2		d [16]			
CF_4/Cl_2			c [20]		
SF_6/Cl_2			c [20]		
SiF_4/Cl_2			c [9]		
SF_6/CCl_4	c [11]				c [11]
NF_3/CCl_4	c [11]				c [11]
SF_6/HCl	c [11]				c [11]
NF_3/HCl	c [11]				c [11]

* Explanation of letters and numbers used:
(a) Barrel plasma etching;
(b) Planar plasma etching;
(c) Reactive ion etching, reactive sputter etching;
(d) Triode etching;
(?) Reference ambiguous on reactor configuration and/or specific gas used.

References:
[1] Mochizuki et al. (1978); [2] Chow and Steckl (1980a); [3] Whitcomb and Jones (1982); [4] Mochizuki et al. (1980); [5] Chow and Steckl (1980b); [6] Chow and Steckl (1982); [7] Okazaki et al. (1981); [8] Beinvogl and Hasler (1981); [9] Zhang et al. (1983); [10] Gorowitz and Saia (1982); [11] Chow and Fanelli (1984); [12] White et al. (1982); [13] Crowder and Zirinsky (1979); [14] Bennett et al. (1981); [15] Coe and Rogers (1982); [16] Ephrath and Bennett (1982); [17] Neppl and Schwabe (1982); [18] Sinha et al. (1980); [19] Light and Bell (1984); [20] Mattausch et al. (1983); [21] Wang et al. (1982); [22] Rude et al. (1982).

I. PLANAR PLASMA ETCHING

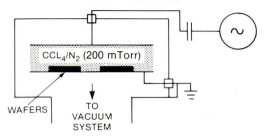

II. REACTIVE ION ETCHING

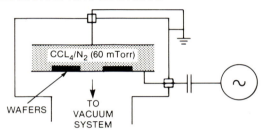

III. SPUTTER ETCHING

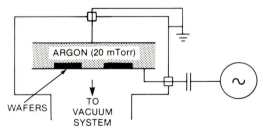

Fig. 2. Schematic diagrams of two plasma etching reactor configurations and conventional sputter etching (from Chow and Steckl, 1983a).

RIE modes is in the r.f. power coupling, with the power supply connected to the top electrode for PPE and to the bottom electrode (on which the wafers are situated) for RIE. The FDE and TE modes are further refinements of the first two. In FDE, the r.f. power can be switched between the two electrodes during the process, while in TE, both of the electrodes are simultaneously powered (with the chamber wall grounded).

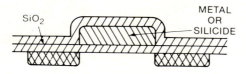

SINGLE LEVEL: METAL OR SILICIDE GATE

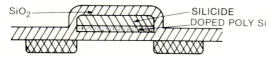

COMPOSITE: SILICIDE/POLY Si GATE (POLYCIDE)

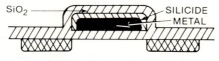

COMPOSITE: SILICIDE/METAL GATE

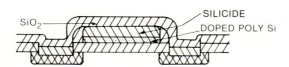

COMPOSITE: SELF-ALIGNED
SILICIDE/POLY Si GATE (SALICIDE)

Fig. 3. Various gate structures incorporating refractory metals and/or metal silicides (from Chow and Steckl, 1983b).

For a more detailed discussion of plasma etcher design and etching modes, see Chapman (1980), Ephrath (1982), and Vossen (1979). Since each type of reactor may only be optimized for the specific application, any direct comparison among these etching modes is difficult and must take into account the system design and operating parameters (such as pressure, electrode spacing). Furthermore, partly due to the complexities of these plasma processes and partly due to the lack of quantitative descriptions of the basic surface chemical reactions, no adequate model exists at present to predict the relative trend of the etching behaviors for specific etchant/refractory materials combinations. Hence, all information obtained has so far been empirical.

Gate structures incorporating refractory metals and/or metal silicides are shown in fig. 3 (Chow and Steckl, 1983b) and discussed in the following sections. Since the patterning of these refractory materials is intricately related to the MOS gate structures being implemented, it is useful to first point out various gate structures that have been suggested and their related etching requirements. The simplest gate structure is a direct replacement of the poly-Si with a refractory metal or metal compound like silicide. The patterning process, in this case, should etch the gate conductor anisotropically and selectively over the gate dielectric (SiO_2, Si_3N_4) and the masking material (photoresist). Despite the simplicity of the single-level gate approach, the reliability of the poly-Si gate has led to the implementation of various composite gates in which the poly-Si/SiO_2 interface is maintained while the conductivity is enhanced with a silicide overlayer. The etching requirements for polycide (silicide on poly-Si) are more demanding. Now both materials must be etched anisotropically and with selectivity to mask and gate insulator and lateral etch rates must be such that a smooth etch profile is produced. In the following sections, the etching of single-level refractory metal and silicide gates as well as polycide and other composite gates is discussed.

2. Pattern definition of single-level gates

2.1. Metals

Among the refractory metals, molybdenum and tungsten are most widely used as MOS gates and interconnections. One of the major conditions for a metal to be considered as a gate material is its stability with the gate dielectric which is usually SiO_2 or Si_3N_4. The group consisting of molybdenum, tungsten and rhenium is the most stable over both oxide and nitride (Chow, 1982; Pretorius et al., 1978) and hence most appropriate for microelectronic applications. Other metals, such as titanium and tantalum, can react with the oxide or nitride at relatively low temperatures (~600°C) and thus may not be placed directly over the gate dielectric. However, these other refractory metals have been used in other applications, such as interconnections in hybrid circuits and superconductors in Josephson Junctions. Therefore, the etching gases and modes used for their patterning have been included in table 4 along with those for molybdenum and tungsten.

2.1.1. Fluorinated plasmas

It has been known for some time now that fluorocarbon plasmas, usually in conjunction with oxygen, can etch many refractory metals. As with silicon etching, many of the initial studies were performed in barrel- or downstream-type reactors with CF_4 without (Maeda and Fujino, 1975; Takahashi et al., 1978; Busta et al., 1979) or with oxygen addition (Legat and Shilling, 1975; Horiike and Shibagaki, 1977; Poulsen, 1977; Chow, 1982). In one barrel etcher, an etch rate of 30 nm/min was obtained for Mo, and adding 4% oxygen to CF_4 enhances the etch rate by about a factor of 4 (fig. 4) (Chow, 1982). Subsequent studies utilized planar-type reactors in either the PPE (Shah, 1979) or RIE mode (Lehmann and Widmer, 1979). Recently, other fluorinated gases, most notably NF_3 and SF_6, which can generate more fluorine atoms or radicals, have also been investigated (Chow and Steckl, 1980b, 1982; Roth and Hagen, 1980; Randall and Wolfe, 1981; Bensaoula et al., 1983), resulting in much higher etch rates. In one study (Chow and Steckl, 1980b, 1982) pure NF_3 as well as NF_3 diluted with argon and helium were used. An etch rate of 200 nm/min was obtained at a pressure of 100 m Torr as shown in fig. 5. A drastic decrease in Mo etch rate was observed when argon or nitrogen was added (fig. 6), probably due to a reduction in reactant concentration. In another study, the tungsten etch rate was found to be monotonically decreasing with increasing oxygen content in a reactive ion etcher with SF_6/O_2 at 10–20 mTorr, as shown in fig. 7. Furthermore, while both molybdenum and tungsten can be selectively etched over SiO_2, the etch rate of Si_3N_4 is usually faster than those of the metals in fluorine-based plasmas (Chow, 1982).

Edge profiles of these metals have only been studied in a few of these plasmas. A vertical-to-horizontal etch ratio of 3 was measured for NF_3 etching of Mo at 100 mTorr in PPE (Chow and Steckl, 1982) (fig. 8) while very anisotropic profiles were obtained for both Mo and W in SF_6/O_2 using RIE at 10–20 mTorr and high power (Randall and Wolfe, 1981; Bensaoula et al., 1983). Also, the side etching rate of Mo was found to be dependent on film thickness in a CF_4 plasma (Takahashi et al., 1978). While it was constant for thicker films, it exhibited a saturation trend when the film became thinner (<300 nm).

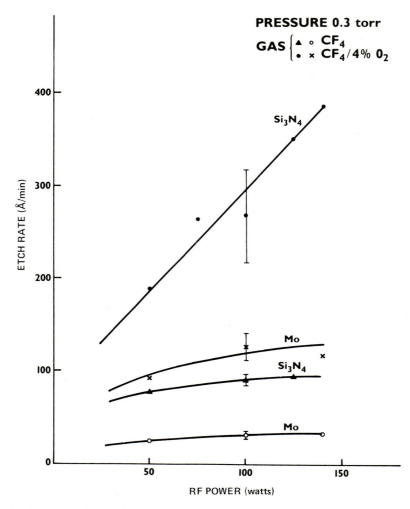

Fig. 4. Etch rates of molybdenum and Si_3N_4 as a function of r.f. power for CF_4 and $CF_4/4\%$ O_2 at 0.3 Torr in a barrel reactor (from Chow, 1982).

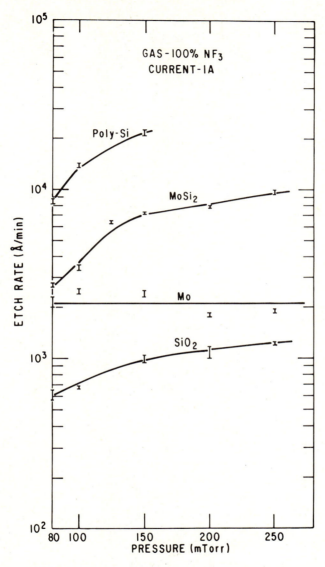

Fig. 5. Etch rates of Mo, MoSi$_2$, doped poly-Si and SiO$_2$ are shown as a function of reactor pressure at 1 A in NF$_3$ (from Chow and Steckl, 1982).

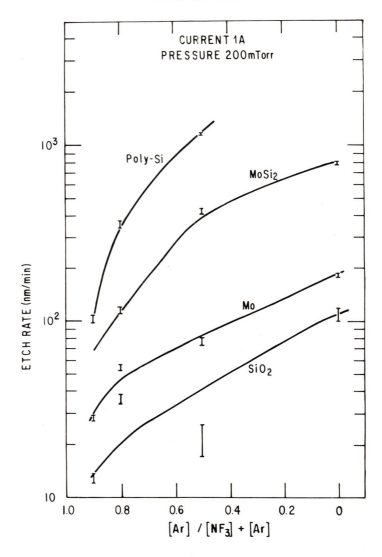

Fig. 6. Etch rate dependence of Mo, MoSi$_2$, doped poly-Si and thermally grown SiO$_2$ on argon percentage in NF$_3$ for a planar plasma etcher (from Chow and Steckl, 1982).

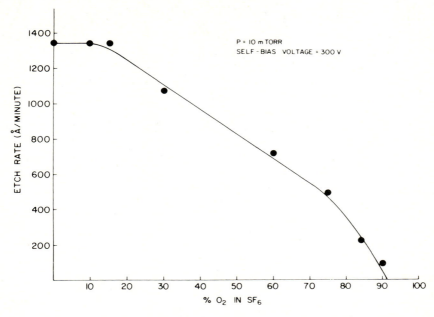

Fig. 7. Etch rate of tungsten as a function of oxygen content in SF_6/O_2 mixtures for a reactive ion etcher (from Randall and Wolfe, 1981).

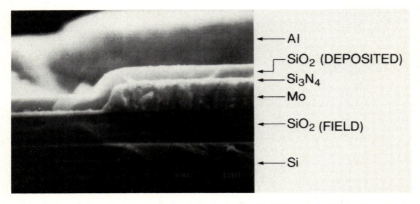

Fig. 8. SEM cross-sectional view of an edge profile of a 300 nm thick Mo film which was etched in NF_3. The resist has been removed and various thin film layers (as indicated) have been deposited over the Mo layer. (From Chow and Steckl, 1982.)

2.1.2. Chlorinated plasmas

Similar to CF_4, CCl_4 has been most thoroughly studied among the chlorine-based compounds. Pure CCl_4 plasmas not only generate much polymers in the etching chamber but also etch Mo and W very slowly and hence cannot be used. Addition of large amounts of oxygen (over 50% by volume) greatly enhances the etch rate and selectivity over SiO_2 (60–100) (Hirata et al., 1981; Kurogi and Kamimura, 1982; Gorowitz and Saia, 1982). The reason for the large increase in etching is believed to be the formation of molybdenum oxychlorides which have a much higher volatility than the chlorides. Obviously, because of the large amounts of oxygen

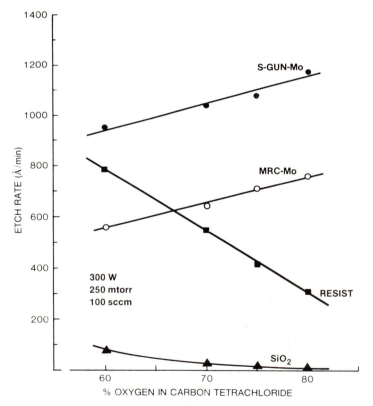

Fig. 9. Etch rate of molybdenum films deposited in two different sputtering systems as a function of oxygen percentage in CCl_4/O_2 mixtures (from Gorowitz and Saia, 1982).

added, the resist etch rate is also increased but this is considered as an acceptable trade-off. For example, when the oxygen percentage is varied from 20 to 80%, the photoresist (AZ1350J) etch rate is only tripled while the Mo etch rate is five times higher (Kurogi and Kamimura, 1982). Similar variations in etch rate with oxygen content were measured for oxygen percentages between 60 and 80% (fig. 9) (Gorowitz and Saia, 1982). A similar dependence on oxygen content was observed for negative electron beam resist (Hirata et al., 1981). In addition, the silicon etch rate was sharply suppressed at over 70% of oxygen, making selective etching of Mo over silicon feasible (Kurogi and Kamimura, 1982). As will be discussed in section 3, similar characteristics are expected for tungsten, rhenium and chromium, but only the case of chromium etching has been verified in the context of photomask making (Nakata et al., 1980; Suzuki and Yamazaki, 1982). Actually, Suzuki and Yamazaki (1982) used carbon dioxide instead of oxygen as the oxidant so as to retard photoresist degradation.

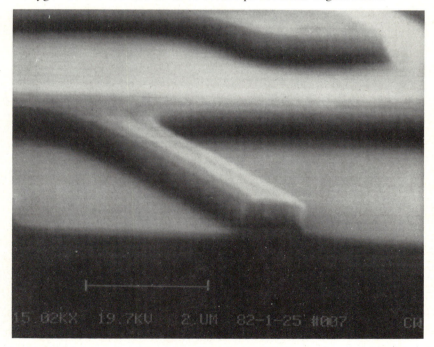

Fig. 10. SEM picture of an Mo edge profile etched in CCl_4/O_2, showing negligible undercutting (from Gorowitz and Saia, 1982).

The edge profiles obtained in CCl_4/O_2 plasmas were almost totally vertical (See fig. 10, Gorowitz and Saia, 1982). This may be because the molybdenum oxychlorides can be formed or desorbed only under energetic ion bombardment. In fact, traces of oxychlorides can be found on the electrode surfaces and the chamber walls after etching. Thus, precautions must be taken to prevent contamination of the wafer surfaces. To clean up the chamber walls after etching, a fluorinated plasma (like CF_4) can be used (Gorowitz and Saia, 1982).

2.1.3. Halide mixtures

To improve the etching characteristics, combinations of fluorine- and chlorine-based gases can be used. Basically, there are two ways of introducing two or more halide components into the etching reactions. The first is direct mixing of the gases at the inlet of the reaction chamber. Since the ratio of each component is controlled by the gas flow, it can be adjusted at will. The second is to use a molecule containing all of the halide components, such as fluorochloromethane.

As far as halide mixtures are concerned, the refractory metals have been mostly patterned using the second approach. In this context, reactive sputter etching of Mo with CCl_2F_2 and $C_2Cl_3F_3$ has been reported in an early study (Hosokawa, 1974). At 20 mTorr, the Mo etch rate was only 18.5 nm/min in an argon plasma but increased to 83.6 and 77.5 nm/min when CCl_2F_2 and $C_2Cl_3F_3$ were used, respectively, as etchants. Also studied has been CF_3Br etching of Mo over SiO_2 and Si_3N_4 with resist as mask in the reactive ion etching mode (Matsuo, 1980a). No undercutting of the etched layer was observed. Also, selectivities of 4–6 and 1.5–2 were respectively obtained for Mo over oxide and nitride at 30 mTorr with fused quartz as the cathode material. PPE of Ti using CF_3Br with He/O_2 addition has also been reported (Mogab and Shanoff, 1977). Helium was found to stabilize and make uniform the discharge while oxygen addition slightly increased the generation of halide species and also prevented organic particulate contamination.

2.2. Metal silicides

The stability trends mentioned earlier for refractory metals on SiO_2 and Si_3N_4 also apply to the metal silicides (Murarka, 1980; Chow, 1982). The situation is less critical for the silicides because among the silicide compounds, only the disilicides which have a large percentage of silicon

(66.7%) are commonly used. The stability over SiO_2 or Si_3N_4 can be further enhanced by putting excess silicon into these metal disilicides. Nevertheless, the silicides most often implemented in single-level structures are still those of molybdenum and tungsten. Etching gases and modes reported so far for this class of materials are listed in table 5 (section 1.3).

2.2.1. Fluorinated plasmas

Similar to the historical development of etching of metals, fluorocarbons were initially used for the silicides. Etch rates of 80–100 nm/min have been reported for $MoSi_2$ in a barrel reactor (Chow and Steckl, 1980a) and $NbSi_2$ (Rude et al., 1982) at 100 W of r.f. power and a pressure of 0.2–0.3 Torr with an etch rate selectivity over thermal oxide of $\sim$ 15 to 1. The etch rate dependence of as-deposited $MoSi_2$ films on applied r.f. power is shown in fig. 11 (Chow and Steckl, 1980a). Isotropic profiles were generally obtained with this reactor/etchant combination. In NF_3 plasmas, the silicide

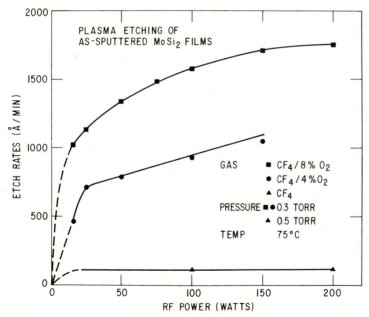

Fig. 11. Etch rate dependence of as-sputtered $MoSi_2$ films on r.f. power for various CF_4/O_2 mixtures in a barrel etcher (from Chow and Steckl, 1980a).

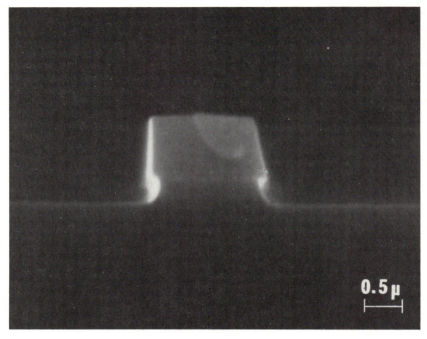

Fig. 12. SEM cross-sectional view of a 1 μm $MoSi_2$ line etched in 100% NF_3 at 1 A and 100 mTorr. The ~1 μm-thick photoresist that was used as mask has not yet been removed. (From Chow and Steckl, 1982b.)

etch rate is generally higher than for CF_4/O_2, due to the more efficient generation of atomic fluorine. While the selectivity (4–8) of $MoSi_2$ over SiO_2 was lower than for the previous case (15–20), more anisotropic profiles were realized as demonstrated in fig. 12 for a 1 μm line. As can be observed from fig. 6, above, diluting NF_3 with inert gases (argon in this case) resulted in similar decrease in the etch rate for $MoSi_2$ as for Mo (Chow and Steckl, 1982).

2.2.2. Chlorinated plasmas

Compared to fluorinated plasmas, fewer works have been reported on chlorinated plasmas. Gorowitz and Saia (1982) performed RIE of $MoSi_2$ using CCl_4/O_2. Generally, lower etch rates of both silicide and oxide were measured than in the NF_3 case. The relationship between etch rates (of

MoSi$_2$, SiO$_2$ and photoresist) and pressure for 50% O$_2$ in CCl$_4$ is shown in fig. 13. The silicide etch rate increases with pressure (as in NF$_3$ plasmas), whereas the oxide etch rate is constant (while it increases with pres-

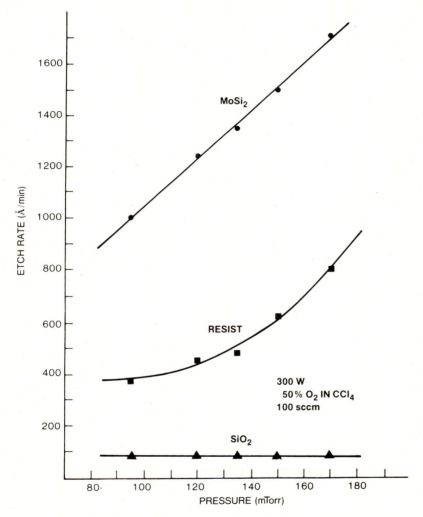

Fig. 13. Etch rate of MoSi$_2$, SiO$_2$ and photoresist at 300 W in 50/50 CCl$_4$/O$_2$ vs. reactor pressure (from Gorowitz and Saia, 1982).

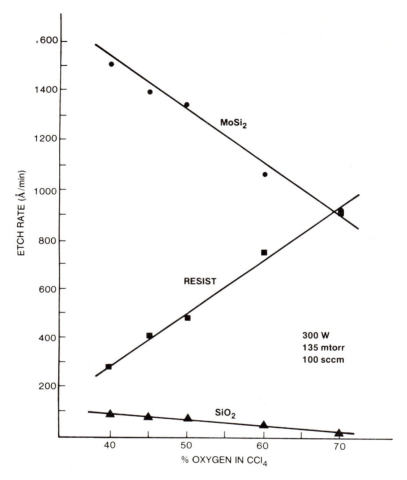

Fig. 14. Etch rates of MoSi$_2$, SiO$_2$ and photoresist films as a function of oxygen percentage in CCl$_4$/O$_2$ mixtures (from Gorowitz and Saia, 1982).

sure in NF$_3$). For the same RIE system, CCl$_4$ diluted with oxygen (40 to 70%) resulted (Gorowitz and Saia, 1982) in a lowered etch rate for both MoSi$_2$ and SiO$_2$ (fig. 14).

2.2.3. Fluorinated/chlorinated mixtures

Recently, several combinations of fluorine/chlorine compounds have been investigated for a few silicides. However, most of the studies addressed the problem of polycide etching and hence concentrated on optimizing silicide/poly-Si etch rate ratio and edge profiles. These results will be discussed later, in section 3.1. Here, we will only discuss the studies concentrating on selective etching of silicides over oxide or nitride. In one recent study, RIE of $MoSi_2$ and $NbSi_2$ was investigated with SF_6 or NF_3/CCl_4 or HCl mixtures. The details will published elsewhere (Chow and Fanelli, 1984). In general, the silicide and poly-Si etch rate increases with increasing fluorine content, r.f. power and pressure. For example, at 100 W in a 50/50 NF_3/HCl mixture, the etch rates of $MoSi_2$ and SiO_2 were 37 and 6 nm/min but were increased to 54 and 30 nm/min respectively when the r.f. power was raised to 300 W. Also, HCl and NF_3 plasmas tend to have a higher substrate d.c. bias (by about an order of magnitude) than CCl_4 or SF_6 plasmas under identical conditions.

2.3. Metal nitrides

Because of their low resistivity ($\sim 50\,\mu\Omega\,cm$) and inertness in chemical solvents used for integrated-circuit processing, metal nitrides have also been investigated as replacement for doped poly-Si. Most notable among them is titanium nitride which has been implemented in discrete MOSFET's (Wittmer et al., 1983) and CMOS circuits (Noser et al., 1982). Patterning of the nitride layer was performed in a CF_4 plasma (Wittmer et al., 1983; Noser et al., 1982) over gate oxide with negative resist masking, but little detail on the process has been given.

3. Pattern definition of composite gates

3.1. Polycide

Polycide is a composite gate consisting of a refractory metal silicide on doped poly-Si. The silicides used include $TiSi_2$ and $TaSi_2$ as well as $MoSi_2$ and WSi_2. The use of $TiSi_2$ and $TaSi_2$ is allowed here since the silicide in a polycide structure does not contact the gate oxide. The thicknesses of the silicide and poly-Si layers are usually between 250 and 300 nm each. The poly-Si is n^+ doped, usually by the decomposition of $POCl_3$. The use of an additional thin film on top of the silicide has been reported, the purpose being to control the out-diffusion of dopant during high temperature anneals.

The use of polycide places a requirement on the etching process beyond those for a single level of poly-Si, metals or silicides. Besides controlling the linewidth and providing sufficient selectivity to SiO_2, the lateral etch rates of both poly-Si and silicide must be such that the etch profile is smooth. A faster lateral etch rate for the doped poly-Si will produce a silicide overhang. Such a structure is undesirable because contaminants may be trapped underneath the silicide and because link-up may not occur between the polycide gate and the ion-implanted source/drain region.

The etch rate ratio that is needed between the poly-Si and the underlying SiO_2 depends in part on the etch rate ratio between the poly-Si and the silicide. As the silicide etch rate falls relative to that of the poly-Si, the minimum etch rate ratio needed between the poly-Si and the underlying SiO_2 increases. Calculations showing the change in minimum poly-Si to SiO_2 etch rate ratio as the poly-Si to silicide etch rate ratio is varied have been made (Wang et al., 1982). Polycide etching has most often been carried out in parallel-plate or radial flow reactors, in either the PPE or RIE mode as well as in the FDE mode.

Etching processes for polycide have been based on fluorinated plasmas (CF_4 and SF_6), chlorinated plasmas (Cl_2 and CCl_4) and on plasmas of freon containing both fluorine and chlorine. Oxygen is often added to the etching gases to increase silicide etch rate as well as to adjust the silicide to poly-Si etch rate ratio. The processes often consist of several steps to obtain an acceptable combination of etch profile and selectivity to SiO_2.

3.1.1. Fluorinated plasmas

Etching of polycides formed with Mo, W and Ta silicides has been described by Beinvogl and Hasler (1981). A two-step RIE process using SF_6/He was developed. Anisotropic etching with low selectivity to SiO_2 is used to etch the silicide plus most of the poly-Si. Etching conditions that produce some undercutting but provide selectivity to SiO_2 of 10 to 1 are used to finish the etch. Etch rates of W, Mo and Ta silicides are reported to be similar but the etch rate of poly-Si under $TaSi_2$ is enhanced.

PPE using mixtures of SF_6 and O_2 have been used to pattern polycide formed with silicon-rich tungsten silicide (Coe and Rogers, 1982). The percentage of oxygen was important in determining the etch rate ratio to SiO_2 with approximately 10 to 1 being obtained in SF_6/35% O_2. The silicide films were deposited silicon-rich because the use of stoichiometric WSi_2 films was associated with severe undercutting in the poly-Si. The films were not annealed before etching because annealing resulted in lateral etching of the Si-rich silicide.

Polycide formed with $TaSi_2$ has been etched using CF_4/O_2 in a parallel plate reactor (Sinha et al., 1980). It is reported that the $TaSi_2$ does not undercut the resist mask but that the poly-Si is etched isotropically, producing a silicide overhang of about 300 nm per edge.

A process for producing micron-dimension WSi_2 polycide features with a sloped etch profile has been reported (Tsai et al., 1980). A controlled slope is advantageous when polycide is used as the first level in a double poly-Si FET process because the overetch required to remove the second level of poly-Si from the sidewalls of the first is reduced. The process is carried out in two steps in a dual electrode (flexible diode) reactor with a mixture of CF_4 and O_2 as the etchants and aluminum patterned by lift-off as the etch mask. The silicide plus one-half the poly-Si is etched in the RIE mode at 100 mTorr. A slope of about 60° is introduced in this step. The remaining poly-Si is then plasma etched with high selectivity to SiO_2 (50 to 1) by operating the reactor in the PPE mode. While in the RIE step the etch rates are relatively insensitive to the percentage of O_2 in CF_4 and to n-type dopant concentration, the plasma etch rate of WSi_2 depends sensitively on these parameters. To adjust and make reproducible the etch rate of WSi_2, it was necessary to deposit a thin layer of SiO_2 on the polycide stack to prevent out-diffusion of phosphorus during anneal. The rate at which the WSi_2 is etched laterally during plasma etching was reduced and made less sensitive to the etching gas composition by reducing the oxygen percentage to 35%. A smooth etch profile with an etch bias of ~ 0.1 μm was obtained.

3.1.2. Chlorinated plasmas

Wang et al. (1982) reported a single-step etching process for $TiSi_2$ polycide using a chlorine-based plasma. Micron-dimension structures are etched anisotropically with selectivity to SiO_2 of 15 to 1. The importance of depositing carbon- and oxygen-free silicide was stressed for clean etching. The polycide is etched after annealing, even though annealing increases the poly-Si to silicide etch rate ratio from 1 to 2, because the unannealed silicide undercuts significantly. The disadvantage to etching after annealing is that the higher poly-Si to silicide etch rate ratio increases the etch rate ratio that is needed to the underlying SiO_2; from 15 to 1 to 20 to 1 in this case. Details of the reactor and the etching mixture were not given.

Mattausch et al. (1983) performed RIE of cosputtered $TaSi_2$ polycide in

SF_6/Cl_2 at ~ 40 mTorr. The composition determines the etch profile with poly-Si undercutting the silicide in pure SF_6 and silicide undercutting relative to the poly-Si in Cl_2 rich mixtures. The optimum profile obtained at $Cl_2/62.5\%$ SF_6 exhibits a 0.2 μm undercut before overetching for a 20 nm thick silicide film. A low bias two-step process in which SF_6/Cl is used to etch into the poly-Si and pure chlorine is used to etch the remainder of the poly-Si is also described. The etch rate ratio to SiO_2 is higher, greater than 50 to 1, for both etching steps. Annealing is not carried out before etching because annealing roughens the $TaSi_2$/poly-Si interface. A smooth interface was reported to be crucial for determining the switching point in the two-step process.

3.1.3. Halide mixtures

A three-step process for patterning submicron lines in $MoSi_2$ polycide has been described (Whitcomb and Jones, 1982). Two of the three steps rely on recombinant chemistry similar to that described elsewhere for etching poly-Si (Mogab and Levinstein, 1980) and aluminum (Hess, 1981). In these procedures, directional etching is obtained by passivating vertical surfaces with a relatively non-volatile species. Etching occurs on horizontal surfaces because these surfaces are kept 'clear' by ion bombardment. Specifically, the $MoSi_2$ is etched in a mixture of CF_4 and C_2F_6 and the poly-Si is etched in CCl_2F_2/C_2F_6. Ion milling is used between these two steps to remove a relatively non-reactive interfacial layer that is about 25 nm thick. The reduction in feature size reported using this three-step process is less than 0.1 μm.

Directional and selective etching of WSi_2 polycide was obtained using CCl_2F_2 in a dual electrode (flexible diode) reactor (Ephrath and Bennett, 1982). Unlike the two-step (RIE followed by plasma etching) process described using this reactor in section 3.1.1 etching is carried out in one step with both electrodes powered, decreasing the d.c. bias simultaneously. Etch rate ratio to SiO_2 is maximized by decreasing the d.c. bias on the substrate electrode to the minimum needed for vertical etching. In this way, an etch rate ratio of 15 to 1 is obtained for poly-Si and WSi_2 to SiO_2. Etching is carried out at a low pressure of 25 mTorr so etching remains directional during extended overetching. CCl_2F_2 is a convenient etching gas because it is uniform and shows little load dependence. Etch rates are insensitive to pattern density and residual oxygen and water vapor.

3.2. Metal-nitride/metal gates

Recently, molybdenum gates overcoated with a layer of its own nitride have been studied to avoid dopant ion channeling. The nitride layer can be formed by reactive sputtering of Mo in Ar/N_2 (Okabayashi et al., 1981) or nitridation of Mo with forming gas (N_2/H_2) or ammonia (Kim et al., 1983). Unless the nitride is formed through self-aligned nitridation after the Mo patterning, patterning of the composite molybdenum nitride/Mo stack is needed. Only limited data have been reported for RIE of Mo_2N in CCl_4/O_2 (Gorowitz and Saia, 1982). The nitride etch rate is generally about 40% higher than the Mo etch rate. For example, etch rates of 146 and 107 nm/min were obtained for Mo_2N and Mo respectively, in 75/25 CCl_4/O_2 at 250 mTorr. Furthermore, lateral undercut of only 0.1 μm on each side was measured for a Mo_2N/Mo composite.

4. Etching dependence on thin film properties

The etching characteristics of refractory metal and silicide films is dependent on thin film properties, for example, stoichiometry, crystallinity, level of contaminants and concentration of dopants. These properties are, in turn, highly dependent on the deposition system parameters and subsequent heat treatment.

Deposition conditions known to affect etching include partial pressure of residual gas, substrate temperature, and deposition rates of silicon and refractory metals which determine the stoichiometry of the film (in the codeposition case). The effect of the residual gas on the etch rate of $TaSi_2$ was studied by introducing N_2, O_2 or H_2O (11.7×10^{-6} partial pressure) during co-evaporation of tantalum and silicon (Neppl and Schwabe, 1982). The etch rate in CF_4 decreased only slightly for the films deposited in the presence of nitrogen or water vapor, but by a factor of two for films deposited in the presence of oxygen. In this same study, it was found that evaporation at a high temperature (300°C) produced higher etch rate films than those deposited at room temperature, presumably because incorporation of impurities was reduced at the higher temperature. The opposite dependence on oxygen level is reported for molybdenum (Gorowitz and Saia, 1982). An enhanced Mo etch rate for films sputtered from a batch (non-load-locked) system (fig. 9, above) is attributed to a higher oxygen level.

Coe and Rogers (1982) found that the stoichiometry of WSi_2 in a polycide structure affects the etch rate of the underlying poly-Si. Polycide structures with a pronounced WSi_2 overhang were produced when the polycide stack was etched in SF_6. Substantial improvement in profile resulted when the cosputtered WSi_2 films were deposited silicon-rich. These films were not annealed before etching.

Annealing conditions known to affect etching are the time and temperature of the anneal as well as the presence of contaminants in the inert ambient. The purpose of the anneal is to crystallize the silicide to reduce its resistivity. An etch rate difference of 15–20% was observed for $MoSi_2$ films before and after annealing, probably due to the more crystalline nature of the annealed films (Chow and Steckl, 1980a). An additional effect of the anneal on polycide films is to redistribute and cause the loss of dopant. Etch rates of poly-Si and silicide tend to increase with the concentrations of n-type dopants and so the amount and distribution of dopants after the anneal directly affects the profile of the etched polycide structure. The diffusion of dopants in WSi_2/poly-Si films has been discussed by Pan et al. (1982) and Jahnel et al. (1982) and the diffusion of phosphorus in $TaSi_2$/poly-Si films has been reported by Maa et al. (1983) and Pelleg and Murarka (1983). The loss of dopant leads to an increase in resistivity and so a cap of oxide is sometimes used to prevent out-diffusion. The annealing ambient contaminant of most concern is oxygen. For example, it was found that WSi_2 grains formed in oxygen-contaminated argon were large and defect-free but the film contained many large voids. WSi_2 grains formed in high purity argon were smaller and contained many defects but were free of voids. The presence of voids could result in poor line definition during etching of 1 µm dimension patterns because the voids are 0.2 µm in diameter and larger.

In another study, it was found that the sheet resistivity of $TaSi_2$/phosphorus-doped poly-Si films was higher after annealing in oxygen contaminated argon than in high purity argon. The difference was attributed to the growth of a thin layer of oxide that served to reduce out-diffusion of the phosphorus (Maa et al., 1983).

5. Endpoint detection

To selectively etch one single layer or multiple layers of thin films over underlying layers or substrates, the ability to detect the endpoint of the plasma etching process is essential. A variety of techniques have been developed to meet this need. Among them are optical emission spectroscopy, mass spectrometry, optical reflection, impedance monitoring, Langmuir probe and pressure change (Marcoux and Foo, 1981). Mass spectrometry and optical emission spectroscopy are apparently the most popular. Both of these techniques involve the monitoring of reactive species and/or etch products. However, the product species are preferred because they originate solely from the etching reactions. For example, in plasma etching of silicon, silicon dioxide and silicon nitride, the peak most often monitored in mass spectrometry is SiF_3^+ ($M/e = 85$ amu) even though other species such as O^+, Si^+ and CO^+ have also been used (Marcoux and Foo, 1981). During the etching of these refractory materials, the product species generally have larger M/e ratios (> 120 amu) than SiF_3^+ and hence are harder to detect in quadruple mass spectrometers. Consequently, the reactant species are preferred but very little work in this area has been reported. Unlike mass spectrometry, optical emission spectroscopy is non-intrusive and generally easier to set up but the identification of the peak signals is sometimes more difficult. For fluorinated plasmas, atomic fluorine and CO (if fluorocarbons are used) are generally monitored for all silicon compounds due to their great sensitivity. Atomic silicon or $SiCl_3$ are usually measured for chlorine-based plasmas. Since the emission lines from either the metals or metal compounds within 200 and 700 nm are often not strong enough for endpoint detection, the reactant species are usually used instead. For the metals, the atomic fluorine line at 704 nm was monitored for CF_4/O_2 etching of LPCVD tungsten on oxide (Marcoux and Foo, 1981) and the atomic chlorine line at 256 nm for RIE of sputtered Mo on oxide with CCl_4/O_2 (Gorowitz and Saia, 1982). In the latter case, a loading effect on the 256 nm-line intensity was observed. During the etching of refractory metal silicides, the peaks used for silicon (F, CO, etc.) can also be used due to the large percentage of silicon present. For example, a photodiode sensitive in the 300–700 nm region has been seen to be effective for PPE of $MoSi_2$ over oxide with CF_4/O_2 in an automated single-wafer reactor (Chow, 1981). Furthermore, because of the high reflectivity of the metals, the optical reflection method which measures the reflectivity changes as the etched layer gets thinner can also yield satisfactory results. In fact, this method was used in conjunction with optical spectroscopy during the plasma etching of tungsten on oxide mentioned earlier as well as etching molybdenum on oxide (Marcoux and Foo, 1981).

6. Mechanisms

To control the etching process and select the proper etchant/metal or silicide combination, one must understand the fundamental mechanisms that occur. While the metals, like silicon, are monocomponent materials, the situation is more complicated for silicides because they are binary compounds. In turn, polycide is an even more complex bilayer structure. Here, we would like to emphasize the distinguishing differences between the etching of these refractory materials and silicon. Several excellent papers are available on the basic reactions that take place during the plasma etching process and their applications for silicon etching (Coburn, 1982; Kay et al., 1980; Flamm et al., 1983). In contrast, basic reactions for the refractory materials have not been investigated.

Surface reactions during etching of metals only need to break the metal–metal bonds, but etching of silicides requires the breaking of metal–silicon and metal–metal bonds as well as silicon–silicon bonds. Also, for binary compounds, there is a greater variety of possible reaction products, the

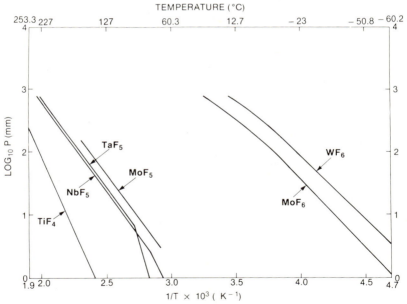

Fig. 15. Vapor pressures of several refractory metal fluorides (from Chow and Steckl, 1983a).

desorption of which can affect the etch rate. The metal halides that are formed in this process are generally less volatile than the silicon halides. Therefore, only the former can be a rate-limiting step in the etching process. In fig. 15, the vapor pressures of some of the pertinent metal fluorides are shown as a function of temperature (Chow and Steckl, 1983a). For comparison, silicon tetrafluoride, which is not included in the figure, has a vapor pressure of over 20 atm at room temperature. WF_6 is the most volatile fluoride in this group while TiF_4 is the least. The observation that titanium residues were found after etching Ti/W with NF_3 (Roth and Hagen, 1980) can be explained by the substantial difference in the volatility of these two metal fluorides. Also, among the metal fluorides, the hexafluorides are more volatile than the pentafluorides and other fluorides. For example, MoF_6 vaporizes at 35°C but MoF_5 boils at 214°C while MoF_4 and MoF_3 are estimated to have higher boiling points. Thus, any lower fluorides formed will remain at the surface until further reactions convert them to the higher fluorides and eventually desorb from the surface. Consequently, the higher metal fluorides are the reaction products expected in the effluent. MoF_6 was detected in CF_4 etching of molybdenum (Dilks and Kay, 1979). TaF_5 was identified as the sole reaction product by modulated molecular beam mass spectrometry in the reaction of molecular fluorine with tantalum between 690 and 930 K at pressures 10^{-5}–10^{-4} Torr in the absence of a plasma (Machiels and Olander, 1979). Also, after partial etching with NF_3 plasma, a $MoSi_2$ film surface was found to be depleted of silicon while molybdenum, fluorine and other species were detected (Chow and Steckl, 1982). This observation indicates that the desorption of metal fluorides could be the rate-limiting step in this case.

The situation is slightly more complicated with the addition of oxygen. In table 6, the melting points, boiling points and applicable sublimation temperatures (some of them are estimated values when experimental data are not available) of silicon and several metal oxyfluorides are shown along with those of the fluorides (Weast, 1978; Glassner, 1959). It can be noted that the oxyfluorides are less volatile than the fluoride compounds. Thus, while on one hand adding oxygen can enhance the generation of fluorine atoms, on the other it can also retard the etching with the formation of a surface layer of metal oxyfluoride. The only metal oxyfluoride detected so far is TaO_2F, formed in the etching of tantalum in CF_4/O_2 plasma (Dybwad, 1983).

For chlorinated plasmas, the addition of oxygen can enhance the etching only when the metal oxychlorides are more volatile than the corresponding

Table 6
Melting, boiling and sublimation points[a] of fluorides and oxyfluorides of several refractory metals and silicon.

Fluoride or oxyfluoride	Melting point (°C)	Boiling point (°C)	Sublimation point (°C)
SiF_4	−90.2	−86	–
Si_2OF_6	−47.8	−23.3	–
TiF_4	> 400	–	284
TiF_3	1200	1400, (1727)	–
TiF_2	(1277)	(2152)	–
ZrF_4	–	–	∼ 600, (903)
ZrF_3	(1327)	(2127)	–
ZrF_2	(1402)	(2277)	–
HfF_4	–	–	(927)
HfF_3	(1327)	(2127)	–
HfF_2	(1377)	(2027)	–
VF_5	19.5	111.2	–
VF_4	–	–	(327)
VF_3	(1127)	(1627)	–
VF_2	(1327)	(2227)	–
VOF_3	300	480	–
NbF_5	72	236	–
TaF_5	96.8	229.5	–
TaF_4	(1477)	(2077)	–
TaF_3	(1237)	(1977)	–
CrF_3	> 1000, 1100	(1427)	1100–1200
CrF_2	1100	> 1300, (2127)	–
MoF_6	17.5	35	–
MoF_5	67	213.6	–
MoF_4	(557)	(617)	–
MoF_3	nonvolatile		
MoO_2F_2	–	–	270
$MoOF_4$	98	180	–
WF_6	2.5	17.5	–
WF_5	(107)	(247)	–
WF_4	(527)	(622)	–
WOF_4	110	187.5	–
ReF_7	48	74	–
ReF_6	18.8	47.6	–
ReF_5	125	221, (387)	–
ReF_4	124.5, (427)	(687)	–
ReF_3	(1107)	(1257)	–
$ReOF_4$	39.7	62.7	–
ReO_2F_2	156	–	–

[a]Values in parentheses are estimated.

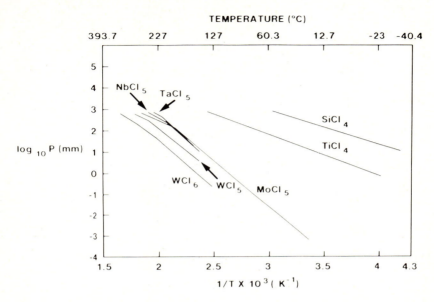

Fig. 16. Vapor pressures of several refractory metal chlorides (from Chow and Steckl, 1983a).

chlorides. The vapor pressures of several metal chlorides are shown in fig. 16 as a function of temperature (Chow and Steckl, 1983a). In contrast to the fluorides, $TiCl_4$ is the most volatile and WCl_6 is the least. The melting and boiling points, and the applicable sublimation temperatures (measured or estimated) (Weast, 1978; Glassner, 1959), of silicon and several metal oxychlorides are shown in table 7. It can be seen that the oxychlorides of Group VIA metals have higher vapor pressure than the corresponding chlorides, whereas the opposite is true for the Group IVA and VA metals as well as silicon. The importance of volatility of oxychlorides has been clearly demonstrated in molybdenum and chromium etching in CCl_4/O_2 plasmas (Hirata et al., 1981; Kurogi and Kamimura, 1982; Gorowitz and Saia, 1982).

7. Summary

Plasma etching of single-level gates of refractory metals, metal silicides and metal nitrides as well as composite polycide and Mo_2N/Mo gates are

Table 7
The melting, boiling and sublimation points[a] of chlorides and oxychlorides of several refractory metals and silicon.

Chloride or oxychloride	Melting point (°C)	Boiling point (°C)	Sublimation point (°C)
$SiCl_4$	−70	57.57	−
Si_2OCl_6	28.1	137	−
$TiCl_4$	−25	136.4	−
$TiCl_3$	(927)	(1327)	−
$TiCl_2$	(757)	(1327)	−
$ZrCl_4$	437[b]	−	331
$ZrCl_3$	(627)	(1207)	−
$ZrCl_2$	(727)	(1387)	−
$HfCl_4$	−	−	319
$HfCl_3$	(627)	(1227)	−
$HfCl_3$	(727)	(1477)	−
VCl_4	−28	148.5	−
VCl_3	dec > 500		
VCl_2	(1000)	(1377)	−
$VOCl$	−	127	−
$VOCl_3$	−77	126.7	−
$NbCl_5$	204.7	254	−
$NbOCl_3$	−	−	335
$TaCl_5$	216	242	−
$TaCl_4$	(297)	(777)	−
$TaCl_3$	(1027)	(1347)	−
$TaCl_2$	(937)	(1377)	−
$CrCl_4$	−	(162)	−
$CrCl_3$	1150	−	947, 1300
$CrCl_2$	824	1302	−
CrO_2Cl_2	−96.5	117	−
$MoCl_6$	(307)	(357)	−
$MoCl_5$	194	268	−
$MoCl_4$	dec 427		
$MoCl_3$	dec		
$MoCl_2$	dec		
$MoOCl_4$	−	−	yes
$MoOCl_3$	−	−	100
MoO_2Cl_2	−	−	yes

Table 7 (cont'd)

Chloride or oxychloride	melting point (°C)	boiling point (°C)	sublimation point (°C)
WCl_6	275	346.7	–
WCl_5	248	275.6	–
WCl_4	dec		
WCl_2	dec		
$WOCl_4$	211	227.5	–
WO_2Cl_2	266	–	–
$ReCl_4$	(177)	500	–
$ReCl_3$	(727)	> 550, (827)	–
ReO_3Cl	4.5	131	–
$ReOCl_4$	29.3	223	–

[a] Values in parentheses are estimated, dec means 'decomposes'.
[b] At 25 atm.

reviewed in the context of patterning micron-size features. Various etching requirements are discussed relative to the gate structures as well as thin film properties. The dependence of etching processes on plasma reactor parameters (such as reactor configuration, pressure and so on) and etchant composition (fluorinated, chlorinated, halide mixtures) is presented. Endpoint detection of these processes is discussed and compared with that developed for silicon and other silicon compounds. The reaction steps distinguishing the plasma etching of this group of materials from that of silicon, particularly the desorption of product species, are discussed.

References

Beinvogl, W., and B. Hasler, 1981, Proc. 4th International Silicon Symposium (The Electrochemical Society, Pennington, NJ) p. 648.
Bennett, R.S., L.M. Ephrath, M.Y. Tsai and C.J. Lucchese, 1981, Electrochem. Soc. Spring Meeting, Extended Abstract **81-1**, 750.
Bensaoula, A., J.C. Wolfe, J.A. Oro and A. Ignatiev, 1983, Appl. Phys. Lett. **42**, 122.
Busta, H.H., R.E. Lajos and D.A. Kiewit, 1979, Solid State Technol. **22** (2), 61.
Chapman, B., 1980, Glow Discharge Processes (Wiley, New York).
Chow, T.P., 1981, unpublished work.
Chow, T.P., 1982, The Development of Refractory Gate Metallization for VLSI, Ph.D. Thesis (Rensselaer Polytechnic Institute).
Chow, T.P., and G.M. Fanelli, 1984, to be published.
Chow, T.P., and A.J. Steckl, 1980a, Appl. Phys. Lett. **37**, 466.

Chow, T.P., and A.J. Steckl, 1980b, IEEE IEDM, Technical Digest, 149.
Chow, T.P., and A.J. Steckl, 1982, J. Appl. Phys. **53**, 5531.
Chow, T.P., and A.J. Steckl, 1983a, A Review of Plasma Etching of Refractory Metal Silicides, Proc. 4th Symposium on Plasma Processing (The Electrochemical Society, Pennington, NJ) p. 362.
Chow, T.P., and A.J. Steckl, 1983b, IEEE Trans. Electron Devices, **ED-30**, 1480.
Coburn, J.W., 1982, Plasma Chem. Plasma Process. **2**, 1.
Coe, M.E., and S.H. Rogers, 1982, Solid State Technol. **25** (8), 79.
Crowder, B.L., and S. Zirinsky, 1979, IEEE Trans. Electron Devices, **ED-26**, 369.
Dilks, A., and E. Kay, 1979, in: Plasma Polymerization, ACS Symposium Series 108, eds. M. Shen and A.T. Bell (American Chemical Society, Washington, D.C.) p. 195.
Dybwad, G.L., 1983, Electrochem. Soc., Spring Meeting, Extended Abstract **83-1**, 324.
Ephrath, L.M., 1982, J. Electrochem. Soc. **129**, 62C.
Ephrath, L.M., and R.S. Bennett, 1982, in: Proc. 1st Int. Symp. on VLSI Science and Technology (The Electrochemical Society, Pennington, NJ) p. 108.
Flamm, D.L., V.M. Donnelly and D.E. Ibbotson, 1983, J. Vac. Sci. Technol. **B1**, 23.
Fukumoto, M., K. Inoue, S. Ogawa, S. Okada and K. Kugimiya, 1981, 1st VLSI Symp. on VLSI Technology, Digest of Technical Papers, p. 28.
Glassner, A., 1959, The Thermochemical Properties of the Oxides, Fluorides, and Chlorides to 2500°K (U.S. Government Report ANL-5750).
Gorowitz, B., and R. Saia, 1982, General Electric TIS Report 82CRD249.
Harada, T., K. Gamo and S. Namba, 1981, Japan. J. Appl. Phys. **20**, 259.
Hess, D.W., 1981, Solid State Technol. **23** (4), 189.
Hirata, K., Y. Ozaki, M. Oda and M. Kimizuka, 1981, IEEE Trans. Electron Devices **ED-28**, 1323.
Horiike, Y., and M. Shibagaki, 1977, Electrochem. Soc. Spring Meeting, Extended Abstract **77-1**, 619.
Hosokawa, N., R. Matsuzaki and T. Asamaki, 1974, Japan. J. Appl. Phys. Supl. 2, Pt. 1, 435.
Jahnel, F., J. Biersack, B.L. Crowder, F.M. d'Heurle, D. Fink, R.D. Issac, C.J. Lucchese and C.S. Petersson, 1982, J. Appl. Phys. **53**, 7372.
Kay, E., J. Coburn and A. Dilks, 1980, Plasma Chemistry of Fluorocarbons as Related to Plasma Etching and Plasma Polymerization, in: Topics in Current Chemistry, Vol. 94, Plasma Chemistry III, eds. S. Veprek and M.. Venugopalan (Springer, Berlin) p. 1.
Kim, M.J., D.M. Brown and W. Katz, 1983, J. Electrochem. Soc. **130**, 1196.
Kurogi, Y., and K. Kamimura, 1982, Japan. J. Appl. Phys. **21**, 168.
Legat, W.H., and H. Schilling, 1975, Electrochem. Soc. Fall Meeting, Extended Abstract **75-2**, 336.
Lehman, H.W., and R. Widmer, 1979, J. Vac. Sci. Technol. **15**, 319.
Light, R.W., and H.B. Bell, 1984, J. Electrochem. Soc. **131**, 459.
Maa, J.-S., C.W. McGee and J.J. O'Neill, 1983, J. Vac. Sci. Technol. **B1**, 1.
Machiels, A., and D.R. Olander, 1979, Electrochem. Soc. Spring Meeting, Extended Abstract **77-1**, 79.
Maeda, K., and K. Fujino, 1975, Denki Kagaku, **43**, 22.
Marcoux, P.J., and P.D. Foo, 1981, Solid State Technol. **24** (4), 115.
Matsuo, S., 1978, Japan. J. Appl. Phys. **17**, 235.
Matsuo, S., 1980a, Appl. Phys. Lett. **36**, 768.
Matsuo, S., 1980b, J. Vac. Sci. Technol. **17**, 587.

Mattausch, H.J., B. Hasler and W. Beinvogl, 1983, J. Vac. Sci. Technol. **B1**, 15.
Mochizuki, T., K. Shibata, T. Inoue and K. Ohuchi, 1978, Japan. J. Appl. Phys. **17**, Suppl. 17-1, 37.
Mochizuki, T., T. Tsujimaru, M. Kashiwagi and Y. Nishi, 1980, IEEE Trans. Electron Devices, **ED-27**, 1431.
Mogab, C.J., and H.J. Levinstein, 1980, J. Vac. Sci. Technol. **17**, 721.
Mogab, C.J., and T.A. Shanoff, 1977, J. Electrochem. Soc. **124**, 1766.
Mogab, C.J., A.C. Adams and D.L. Flamm, 1978, J. Appl. Phys. **49**, 3796.
Murarka, S.P., 1980, J. Vac. Sci. Technol. **17**, 775.
Nakata, H., K. Nishioka and H. Abe, 1980, J. Vac. Sci. Technol. **17**, 1351.
Neppl, F., and U. Schwabe, 1982, IEEE Trans. Electron Devices, **ED-29**, 508.
Noser, J.R., M.C. Jaray and H. Melchior, 1982, Proc. 7th Symp. on Solid State Device Technology, 12th European Solid State Device Research Conference (ESSDERC 82) p. 159.
Okabayaski, H., K. Higuchi and T. Nozaki, 1981, Electrochem. Soc. Spring Meeting, Extended Abstract, **81-1**, 753.
Okazaki, S., T.P. Chow and A.J. Steckl, 1981, IEEE Trans. Electron Devices **ED-28**, 1364.
Pan, P., N. Hsieh, H.J. Geipel, Jr. and G.J. Slusser, 1982, J. Appl. Phys. **53**, 3059.
Pelleg, J., and S.P. Murarka, 1983, J. Appl. Phys. **54**, 1337.
Poulsen, R.G., 1977, J. Vac. Sci. Technol. **14**, 266.
Pretorius, P., J.M. Harris and M.-A. Nicolet, 1978, Solid State Electron. **21**, 667.
Randall, J.N., and J.C. Wolfe, 1981, Appl. Phys. Lett. **39**, 742.
Roth, L.B., and G. Hagen, 1980, unpublished work.
Rude, C.D., T.P. Chow and A.J. Steckl, 1982, J. Appl. Phys. **53**, 5703.
Saraswat, K.C., and F. Mohammadi, 1982, IEEE Trans. Electron Devices **ED-29**, 645.
Saxena, A.N., and D. Pramanik, 1983, Multilevel Metallizations in VLSI, presented at the Gould Electronics Society Symposium, Boston, May 9-11.
Shah, P.L., 1979, IEEE Trans. Electron Devices, **ED-26**, 631.
Sinha, A.K., W.S. Lindenburger, D.B. Fraser, S.P. Murarka and E.N. Fuls, 1980, IEEE Trans. Electron Devices **ED-27**, 1425.
Sinha, A.K., J.A. Cooper, Jr. and H.J. Levinstein, 1982, IEEE Electron Device Lett. **EDL-3**, 90.
Suzuki, Y., and T. Yamazaki, 1982, Electrochem. Soc. Fall Meeting, Extended Abstract **82-2**, 287.
Takahashi, S., F. Murai and H. Kodera, 1978, IEEE Trans. Electron Devices **ED-25**, 1213.
Tsai, M.Y., H.H. Chao, L.M. Ephrath, B.L. Crowder, A. Cramer, R.S. Bennett, C.J. Lucchese and M.R. Wordeman, 1980, J. Electrochem. Soc. **128**, 2207.
Vossen, J.L., 1979, J. Electrochem. Soc. **126**, 319.
Wang, K.L., T.C. Holloway, R.F. Pinizzoto, Z.P. Sobczak, W.R. Hunter and A.F. Tasch, Jr., 1982, IEEE Trans. Electron Devices **ED-29**, 547.
Weast, R.C., ed., 1978, CRC Handbook of Chemistry and Physics, 60th Ed. (CRC Press, Boca Raton, FL) B51–144.
Whitcomb, E.C., and A.B. Jones, 1982, Solid State Technol. **25**, No. 4, 121.
White, F.R., C.W. Koburger, D.L. Harmon and H.J. Geipel, 1982, J. Electrochem. Soc. **129**, 1330.
Wittmer, M., J.R. Noser and H. Melchior, 1983, J. Appl. Phys. **54**, 1423.
Zhang, M., J.Z. Li, I. Adesida and E.D. Wolf, 1983, J. Vac. Sci. Technol. **B1**, 1037.

CHAPTER 3

DRY ETCHING OF GROUP III–GROUP V COMPOUND SEMICONDUCTORS

RANDOLPH H. BURTON, RICHARD A. GOTTSCHO AND GERALD SMOLINSKY

AT&T Bell Laboratories
Murray Hill, New Jersey 07974, USA

Dry Etching for Microelectronics, edited by R.A. Powell
© *Elsevier Science Publishers B.V., 1984*

Contents

1. Introduction — 81
2. Shaping requirements — 82
   2.1. Macroscopic feature etching — 82
      2.1.1. Via hole etching — 85
      2.1.2. Chip separation — 86
      2.1.3. Selective etching and other applications — 88
   2.2. Optical surface etching — 89
      2.2.1. Mirror facet and window formation — 89
   2.3. Etching of surfaces for electronic applications — 92
3. Mechanisms of III–V plasma etching — 94
   3.1. Unifying concepts — 94
      3.1.1. Ion-neutral reactant synergism — 95
      3.1.2. Feedstock composition — 95
   3.2. III–V plasma etching: importance of group III product desorption — 96
      3.2.1. Product desorption — 96
      3.2.2. Ion-enhanced etching and profile control — 103
      3.2.3. Surface roughening and native oxide removal — 105
      3.2.4. Reactants — 106
4. Summary — 107
References — 109

1. Introduction

Semiconducting alloys of group III and group V elements of the Periodic Table have properties which set them apart from other semiconductors and make them suitable for special electronic and optical applications (Mandal, 1982; Bar-Lev, 1979). For example, a high electron mobility ($\sim 8500\,cm^2/$ Vs at 300°C; Sze, 1981) and high useful temperature range (maximum $\sim$ 400°C; Mandal, 1982) make GaAs well-suited for microwave applications (Hindin and Posa, 1982; Mandal, 1982; Bar-Lev, 1979). The direct bandgap of many III–V alloys makes radiative recombination of electrons and holes the most likely minority carrier loss mechanism; this allows forward-biased diodes to perform as efficient light sources (Bergh and Dean, 1976). High drift velocities in an electric field (Sze, 1981), sharp absorption edges, and high absorption coefficients (typical of direct bandgap semiconductors), make many III–V compounds useful as infrared photodetectors. The ability to 'tune' the bandgap by adjusting the alloy composition during crystal growth allows optimization of the material for the intended application.

Unfortunately, III–V compounds present special processing problems. In particular, the diverse chemistries of the group III and group V elements and the great difference in the physical properties of their compounds with other elements make optimization of any etching process difficult. Optimization is further complicated by the fact that many applications impose unique requirements on the etching process. Because of this, developments for each application have proceeded relatively independently of each other and with varying degrees of success.

In this chapter, we review these individual developments with emphasis on the use of plasma or reactive-ion etching for III–V material shaping. In section 2, applications are categorized according to their shaping requirements, and the use of dry etching is reviewed and compared to alternative shaping techniques. In section 3, III–V dry etching mechanisms are discussed, and in section 4, the utility and limitations of III–V dry etching are summarized and areas requiring additional work are defined.

Table 1
Applications requiring the shaping of III–V semiconductors.

Category	Requirements	Applications
Macroscopic features	Fast, material selective, and anisotropic etching	1. Via holes for through-the-chip interconnections 2. Chip separation 3. Wells selectively etched through one layer of a heterostructure
Optical surfaces	Optically smooth surfaces with respect to light propagating in the semiconductor, and well controlled etched profiles	1. Mirror facets or windows 2. Integral lenses 3. Etched gratings
Electronic surfaces	Minimum surface damage, composition changes, or contamination	1. Mesa formation for restricting p–n junction area 2. Channel etching

2. Shaping requirements

Applications requiring the shaping of III–V semiconductors can be separated into three basic categories, those with: (1) macroscopic features requiring fast, selective, and anisotropic etching; (2) optical quality surfaces with well controlled shapes; and (3) channels or mesas with optimized electronic surface properties. Table 1 summarizes III–V device processing goals according to these categories. Since the shaping requirements for each category differ significantly, it is unreasonable to expect a single process to be optimal for all. In fact, the optimal processing conditions for one application often create problems in another application. Similarly, the shaping requirements for these three categories differ markedly from those that characterize most silicon etching applications (i.e. the selective etching of fine lines in thin films). Because of these differences, there has been little technology transfer between Si and III–V applications or even between III–V categories.

2.1. Macroscopic feature etching

Most of the structural designs of III–V devices are fundamentally different from those of silicon devices, even when the devices are functionally similar. Generally, III–V devices are thin (100–200 μm) and are processed on both sides, whereas silicon devices are frequently two to four times thicker and processed on a single side. Compromises in the processing requirements are responsible for these differences. For example, it is desirable to use single side processing because of its simplicity, subsequent ease of access, and reduced packaging costs. With silicon, this is possible because single crystals with low doping densities can be grown in bulk and tailored by ion implantation or diffusion. With III–V materials, however, the desire to vary the alloy composition (for example, in forming sharply defined heterojunctions) and doping levels in a well-controlled manner has dictated that the entire structure be grown in a layered fashion prior to processing. This is illustrated in figs. 1, 2, and 3 which show the structures of a GaAs field effect transistor (FET), an InGaAs/InP p–i–n photodiode, and an InGaAsP/InP double-heterostructure light emitting diode (LED), respectively. The upshot is that functionally important layers exist on opposite sides of the semiconductor and that these wafers must be processed on both sides.

To reduce yield losses from breakage, a wafer should be as thick as pos-

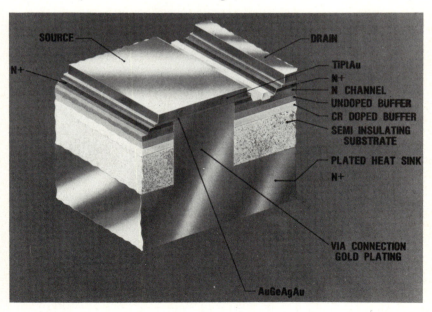

Fig. 1. Artist's drawing of a GaAs field-effect transistor with a through-the-chip (via) interconnection (courtesy of L.A. D'Asaro, AT&T Bell Laboratories).

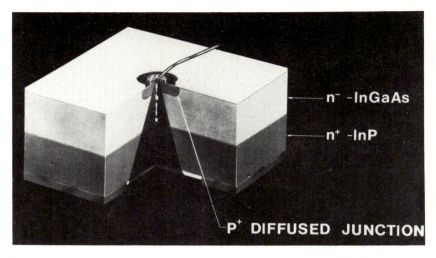

Fig. 2. Artist's drawing of an InGaAs p–i–n photodiode (courtesy of J.R. Zuber and M.A. Pollack, AT&T Bell Laboratories).

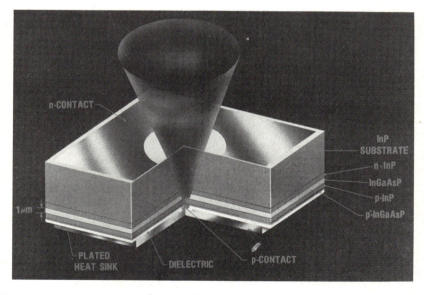

Fig. 3. Artist's drawing of an InGaAsP double-heterostructure surface-emitting LED (courtesy of R.H. Saul, AT&T Bell Laboratories).

sible. Since silicon devices are single sided, their thickness can be optimized for handling. However, performance considerations often restrict the allowable thickness range for III–V devices. For example, thin substrates are desirable for efficient coupling of light from surface-emitting LEDs to optical fibers and from fibers to photodetectors.

These peculiarities of structural design have a direct impact on processing. Since the devices are dual sided, it can be advantageous to etch through a wafer rather than having to access the other side in packaging. In addition, etching can be of value in chip separation. Since the wafers are thin, these approaches are practical.

2.1.1. Via hole etching
If the device structure will allow it, the potential advantages to packaging (and for some devices to performance) are significant enough to warrant through-the-chip interconnections. This was demonstrated for a GaAs FET (see fig. 1) by D'Asaro et al. (1981). They realized that a higher density of devices could be processed on a single wafer if through-the-chip (via) connections of sufficiently small area could be made. They also determined that via transistors would exhibit a higher gain than non-via transistors when operated at frequencies above 4 GHz. However, because of the limitation on area (and therefore undercutting), any etching technique used for forming the via holes must be highly anisotropic. This precludes the use of wet chemical etching.

D'Asaro et al. used plasma etching to fabricate the via holes. They found that with the wafer (at 5°C) mounted on the r.f. driven electrode (at 13.56 MHz, 0.6 W/cm^2) of a parallel-plate radial-flow etcher and a gas mixture of 6% Cl_2 and 94% BCl_3 (50 sccm, 0.15 Torr), they could etch through 30 μm of GaAs (at 0.35 μm/min) in 90 min (including overetch) with only a few μm of undercut and negligible attack of the gold source pad under the hole. Negative photoresist, which etched at 120–150 Å/min under these conditions, was used as the mask.

The two most critical parameters in determining the viability of the technique were the Cl_2/BCl_3 feed gas ratio and the total pressure. As either of these two parameters increased so did both the horizontal and vertical etching rates. Since it was desirable to maximize the vertical etching rate (to minimize processing time) while keeping the anisotropy at an acceptable level, a compromise was struck at the operating conditions mentioned above. Besides the obvious significance to the application at hand, this work is especially notable because it appears to be the only III–V plasma etching process that is routinely used in device production.

2.1.2. Chip separation

A major shaping problem resulting from the thin structures typical of III–V wafers is separation into individual chips without introducing unacceptable levels of mechanical damage. Cracks or defects can readily propagate into the semiconductor and adversely affect both initial yield and long-term reliability. Two common techniques for separation are (1) cleaving, which is capable of forming parallel facets but is difficult to carry out and suffers from low yield, and (2) saw cutting, which is relatively simple and low-cost but results in losses due to damage. A more attractive alternative would be an etching technique, provided that it can be made relatively simple and cost competitive. Generally, wet chemical etching has not been applicable since its isotropic nature causes too much material to be removed laterally to allow efficient use of wafer space (a *de facto* yield loss). The only exception to this would be if the crystal orientation dependence of a wet etchant could be utilized to minimize undercutting.

In 1980, Burton et al. developed a fast, anisotropic plasma etching technique for separating wafers of InGaAsP/InP LEDs (see figs. 3 and 4).

Fig. 4. Scanning electron micrograph of an InGaAsP double-heterostructure surface-emitting LED separated by plasma etching (from Burton et al., 1980).

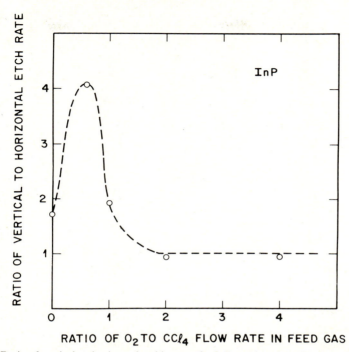

Fig. 5. Ratio of vertical-to-horizontal etching rate for InP as a function of O_2/CCl_4 flow ratio in the feed gas of a 55 kHz discharge (0.05 Torr, 0.23 W/cm^2) (from Burton and Smolinsky, 1982).

They found that with a 1 : 1 mixture of CCl_4 and O_2 (~ 40 sccm, 0.15 Torr) in a parallel-plate radial flow reactor (operated at 55 kHz, 0.15 W/cm^2) and a plate temperature of 300°C, they could etch through a 75 μm thick wafer in about 2h (at 0.6 μm/min). Yields were 100% and the vertical to horizontal etching rate ratio (γ) was 2 : 1. Later, the etching rate and γ were improved to 2.1 μm/min and 4 : 1, respectively, by making adjustments in the gas flow ratio (~ 40% O_2), pressure (0.05 Torr), and r.f. power density (0.23 W/cm^2) (Burton and Smolinsky, 1982). Because of the high temperatures employed, SiO_2 was used as the etching mask.

Two parameters were critical to the success of this scheme: temperature and O_2/CCl_4 flow ratio. In contrast to the work of D'Asaro et al. (1981) on GaAs etching (at 5°C), higher temperatures were found to be necessary for InP in order to insure that the less volatile indium halides could be removed and the native oxide could be penetrated. Burton and Smolinsky

(1982) saw no evidence of etching at temperatures less than 270°C. Figure 5 shows the effects of O_2/CCl_4 flow ratio on the anisotropy of the process. Both of these issues, temperature and the effects of added oxygen, will be discussed further in section 3.

Unfortunately, the plasma separation process suffers from several disadvantages: (1) the need for a SiO_2 mask significantly increases the number of required processing steps; (2) although plasma etching is a batch process, the overall equipment cost makes it relatively expensive; and (3) the 300°C operating temperature, coupled with the relatively long exposure time, can be detrimental to some metallization schemes. In contrast, a relatively new anisotropic etching technique (photoelectrochemical etching), which uses light to induce an anodic decomposition of the semiconductor, appears to be a promising alternative since it is maskless, simple to implement, low-cost, low-temperature and high-yield (Ostermayer et al., 1982; Kohl et al., 1983). Barring significant new developments, a plasma process does not appear to be competitive as a chip separation technique.

2.1.3. Selective etching and other applications

Although one can envision many applications where plasma etching could be utilized to shape macroscopic features, only via hole formation and chip separation have actually been demonstrated on device wafers. This is partly because some applications have been replaced by other technologies. For example, in 1971, Burrus and Miller described a surface emitting GaAlAs/GaAs heterostructure LED that employed a well etched through the GaAs substrate, to eliminate unwanted light absorption. This design also allowed an optical fiber to be located as close as possible to the light emitting region within the LED. However, it required that the etching process be selective for GaAs over GaAlAs. Wet chemical etching can be used to achieve this, although in a confined environment (e.g. a well) bubble formation can cause reproducibility problems. In 1977, Burstell et al. reported a plasma process for selectively etching the GaAs. By using a mixture of CCl_2F_2 and O_2 they were able to achieve rapid GaAs etching ($\sim 10\,\mu m/min$) without attacking GaAlAs. Later, Hikosaka et al. (1981) were able to attain similar selectivities with mixtures of CCl_2F_2 and He. However, other processing developments permitted the Burrus LED to be replaced by a thinner structure in which the GaAs substrate was completely removed (Keramidas et al., 1980). This precluded the use of the plasma process for this application, although its development may not have been for nought, since as devices become more complicated future needs for such a process may arise.

At the moment the scorecard on the use of plasma etching to solve macroscopic III–V shaping problems is mixed. In the three major problems to which it has been applied, it has been successful in one, unable to compete with rival technologies in another, and solved a problem which no longer exists in the third. Treating the latter as a demonstration of potential, the score is two successes to one failure.

2.2. Optical surface etching

One of the most active areas of research involving III–V semiconductors is the development of light sources and detectors for optical fiber communication. By necessity, these devices must be designed to facilitate light transfer into or out of the semiconductor. To be effective, any shaping process employed to form optical surfaces must meet stringent requirements for smoothness and profile control. There is no analogous processing constraint in Si technology.

There are three basic types of optical structures which might be required on III–V devices: (1) flat surfaces which function as windows or mirrors; (2) curved surfaces which act as lenses; and (3) groove arrays which constitute diffraction gratings. The first type of structure has received the greatest attention and will be discussed in the remainder of this subsection. The second type requires that the etching rate be precisely controlled spatially across the sample, a capability plasma etching does not possess. Here, the technique of choice appears to be photoelectrochemical etching, with excellent results reported for lens formation on InP/InGaAsP 1.3 µm double-heterostructure LEDs (Ostermayer et al., 1982). The third application has received relatively little attention from the plasma processing community, although the potential for its use has been demonstrated (Hu and Howard, 1980). This lack of attention is partly due to the fact that a reactive process is not essential for shallow etching; ion milling may suffice. In addition, recent developments using photoelectrochemical etching and interfering laser beams have shown that direct (i.e. maskless) pattern formation with extremely high resolution is possible (Podlesnik et al., 1982).

2.2.1. Mirror facet and window formation

The traditional method for forming flat walls on III–V devices has been scribing and cleaving. This has the advantages of providing smooth optical surfaces and sufficient parallelism to enable lasing. It has the disadvantages of being a low-yield process and of forcing the chip size to the length of the cavity for laser structures. The latter prevents the production of large-area integrated devices and places a lower limit on the cavity length since the chips would become too small to handle or cleave. Because of these disadvantages, there is impetus for developing an etching alternative.

Two tacks may be taken to etch a vertical facet: direct formation or exploitation of a crystal orientation dependence to expose a particular plane. Since the latter is difficult to achieve on a multilayered structure, due to differing material behaviors (all current laser and edge-emitting LED designs employ heterostructures), and is most effective when the plane is already roughly exposed to the etchant, our discussion will concentrate on direct formation. Later we will consider the use of follow-up etches for enhancing the planarity and parallelism of etched facets.

The design rules for developing a technique to form vertical facets are significantly different from those presented in the previous section (2.1) on macroscopic etching. There, we saw that anisotropy, though required, could be traded for etching speed. Here, however, anisotropy cannot be compromised and etching rate is less important (required etched depths extend up to 10 μm). Despite this, standard wet chemical etching has been used (see for example, Iga and Miller, 1982 and references therein), although the results have been both marginal (due to curved walls) and irreproducible (Coldren et al., 1982a). These design considerations lend themselves more to plasma or reactive-ion etching. Since the etching rate is not critical, it is possible to utilize low pressures and high fields to optimize ion bombardment and therefore anisotropy. This approach was taken by Coldren et al., (1980, 1981), who used reactive-ion etching to form mirrors on GaInAsP/InP lasers (see fig. 6) and to form monolithic two-section GaInAsP active-optical-resonator devices.

The conditions used by Coldren et al. were a gas mixture of 80% Cl_2 and 20% O_2, total pressure of 0.002 Torr, substrate temperature of 200°C, r.f. frequency of 13.56 MHz, and r.f. power density of 0.3 W/cm^2. The reactor used was a conventional parallel-plate sputter etching system with the wafers placed on the driven electrode. Although all of these parameters were important to the overall performance of the technique, two, pressure and temperature, were found to be critical. At high pressures and low temperatures, smooth but positively sloped (overcut, see fig. 7) etched walls were obtained; at low pressures and high temperatures, the etched walls were relatively rough but more vertical. This dichotomy, along with

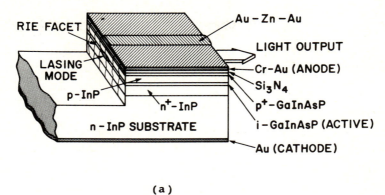

(a)

(b)

Fig. 6. (a) Schematic of a GaInAsP/InP laser with a reactive-ion etched mirror facet. (b) Scanning electron micrograph of the etched facet at the laser stripe. (From Coldren et al., 1980.)

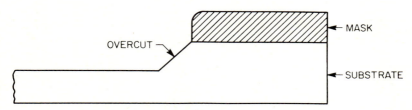

Fig. 7. Schematic drawing illustrating a positively sloped or etched wall.

transfer of mask defects as striations in the wall profile, resulted in non-ideal etched mirrors.

To resolve this conflict and reduce the sensitivity to mask imperfections, Coldren et al. developed a new reactive-ion etching technique which employs both oblique angles of incidence to control wall tilt and sample rotation (or rocking) to wash out surface roughness (Bosch et al., 1981; Coldren and Rentschler, 1981). This technique was then followed by wet chemical etching to obtain the maximum possible facet planarity and smoothness (Coldren et al., 1982a, b) and to achieve performance levels comparable to those of cleaved devices.

In general, the concept of first creating the facet with some of the desired characteristics by dry etching and then finely honing it by wet etching is a sound one. This combination takes advantage of the best characteristics of each technology. While reactive-ion etching is not the only technique available for forming the initial profile (certain combinations of wet etchants may also suffice), it does have the distinct advantage of high final aspect ratios (Coldren et al., 1982a). Note that this advantage is retained after honing since the follow-up etch does not add significantly to the undercut. It is likely that this technology will have an impact on the next generation of optical devices by enabling monolithic integration of lasers or edge-emitting LEDs with other components without sacrifices in performance.

We can now add to our scorecard on III–V dry etching. Of the three applications requiring the etching of optical surfaces (see table 1), mirror formation by dry etching has been clearly successful, integral lens formation has not been possible, and grating formation has not been pursued seriously. Counting the latter two as failures because of an inability to compete with rival technologies, the total score reads three successes to three failures.

2.3. *Etching of surfaces for electronic applications*

In some applications (see table 1), the electronic properties of the etched surface are critical to device performance. If lattice damage, compositional change, or contamination occurs upon etching, then a potentially catastrophic density of surface states can result. For example, drift velocities could be reduced in the near surface region, which could destroy an FET; leakage currents across an exposed junction could increase the dark noise, and therefore reduce the sensitivity, of an optical receiver; and non-radiative recombination of carriers could decrease the efficiency of an LED. In addition, defects at the surface can multiply during subsequent crystal growth, thereby reducing initial yield; or can propagate into the bulk with time, thereby reducing reliability.

The aforementioned concerns place a severe limitation on the types of etching techniques which may be employed. Since none of the available etching techniques is ideal, the least damaging one must be chosen to solve a particular shaping problem. Unfortunately, plasma processes are among the most damaging. By their very nature, they expose the sample surface to a barrage of high energy particles, high temperatures*, redeposition of products (and sputtered electrode or mask materials), and in some cases, polymer deposition. For these reasons, plasma processes have not been used to etch channels or mesas when the electrical properties of the surface are important.

To date, only one research group has demonstrated that moderately low trap densities ($< 10^{15}\,\text{cm}^{-3}$) could be obtained after etching with a plasma-like process (Geis et al., 1981; Lincoln et al., 1982). They utilized a dual beam of Cl_2 molecules and Ar^+ ions to etch GaAs and showed that the resultant surface was suitable for forming high quality Schottky barriers.

Although these results are encouraging, it is unlikely that a dry etching process will supplant the less damaging wet chemical alternatives. The only possible exception to this would be if anisotropy were of primary concern. Given the recent developments in photoelectrochemical etching (Ostermayer et al., 1982, Kohl et al., 1983), even this is unlikely. Therefore, it is our opinion that plasma etching is unable to compete with other processes in the fabrication of device surfaces where electronic performance is critical.

We have now examined a total of eight applications requiring the shaping of III–V semiconductors (see table 1) and plasma etching was found to be the technique of choice for only three. This leaves us with a final score of three successes to five failures. While these results clearly show that plasma etching of III–V is not the cure-all it was for Si, they do provide encouragement for future development.

---

* Even cooled samples experience surface heating from the discharge due to poor thermal conduction by the semiconductor (Schwartz and Schaible, 1981).

3. Mechanisms of III–V plasma etching

In the previous section, we have identified the major problems encountered in dry etching of III–V materials: profile control, surface roughness, surface contamination, and surface states. Understanding the mechanisms of III–V plasma etching can only help to overcome these difficulties and to extend the utility of dry etching techniques. Although much of what is known from etching other materials, such as Si and Al, can be extended directly to etching III–V materials, our knowledge is still insufficient. The major difference between etching Si or Al and etching III–V compounds is that the latter are comprised of elements with diverse chemistries.

We begin this section by reviewing briefly some of the unifying concepts employed in plasma process design; more extensive reviews can be found elsewhere (Mucha and Hess, 1983; Coburn, 1982; Flamm and Donelly, 1981; Coburn and Winters, 1979a; Donelly and Flamm, 1981). We conclude this section by focussing on specific experiments which have elucidated rate-limiting steps in III–V plasma processing and by examining in particular the effects that the substrate's binary composition has on etching rates and surface morphology.

3.1. Unifying concepts

It is convenient to think of plasma etching as proceeding in five steps (Coburn and Winters, 1979a): (1) reactant production in the plasma; (2) reactant transport to the surface; (3) reactant adsorption; (4) product formation; and (5) product desorption. Identifying a microscopic plasma mechanism, therefore, entails determination of the reactants and products as well as the chemical and physical processes by which they are transported, adsorbed, and desorbed.

The difficulties in determining plasma mechanisms stem from the large number of competing reactions associated with each of the five steps above. The degree to which a particular mechanism is important depends on the discharge operating conditions (i.e., power, frequency, pressure, residence time, electrode material, etc.) as well as on the chemical nature of the reactant gas and substrate. For example, when etched depths are less than ~ 100 μm a reaction which is strongly ion assisted may yield isotropic etched profiles if E_s/n, the sheath electric field divided by the total gas density, is not sufficiently large to produce anisotropic ion transport (Zarowin, 1982). Under such conditions it might be falsely concluded that the reaction mechanism was not sensitive to ion bombardment but was purely 'chemical' in nature.

3.1.1. Ion-neutral reactant synergism

One of the most important concepts needed to understand plasma-surface chemistry is the synergism of ion and neutral reactions, because this can affect reactant adsorption, product formation, and product desorption. Coburn and Winters (1979a, b) conclusively demonstrated this synergism by alternately exposing Si surfaces to Ar^+ and XeF_2 beams; the etching rate with both beams on was greater than the sum of the individual rates. It is because of this ion-enhancement of neutral chemistry that anisotropic etching and therefore, very large-scale integration (VLSI) are possible.

Basically two explanations for ion-enhanced surface chemistry have been proposed (Coburn and Winters, 1979a; Donnelly and Flamm, 1981; Donnelly et al., 1983; Mucha and Hess, 1983; Coburn, 1982; Flamm and Donnelly, 1981). First, ion bombardment can stimulate surface reactions; and secondly, ions may stimulate desorption or clear the surface of etch-inhibiting, involatile residues. An ion-stimulated surface reaction might be responsible for the variation in anisotropy observed when InP is etched in low frequency (55 kHz), CCl_4/O_2 plasmas (see fig. 5 and section 2.1.2). This behavior is explained (Burton and Smolinsky, 1982) by competition between: (1) ion-assisted oxidation of surface chlorocarbons which generates surface chlorine and thereby enhances anisotropy; and (2) non ion-assisted production of gaseous or surfaces chlorine which promotes isotropic etching. A surface clearing mechanism may be responsible for the anisotropic etching of GaAs in BCl_3/Cl_2 plasmas (D'Asaro et al., 1981). Apparently, the BCl_3 is a source of contamination which inhibits Cl atoms from chemically etching the side-walls; anisotropic ion transport then serves to keep the horizontal surfaces clean so that vertical profiles are obtained (Zarowin, 1982; Flamm and Donnelly, 1981). In addition, we shall see in section 3.2.2 how ion-assisted product desorption often plays an important role in III–V etching. In general, for anisotropic etching to occur, at least one of these mechanisms must be important and ion transport to the surface must be anisotropic (Zarowin, 1982).

3.1.2. Feedstock composition

Feedstock composition influences the relative importance of each of the five steps because it affects reactivity, contamination, selectivity, and profile control. For example, addition of oxygen to halocarbon discharges affects reactivity and contamination by liberating halogen atoms and reducing polymer deposition (Mucha and Hess, 1983; Burton and Smolinsky, 1982; Coburn, 1982; Flamm and Donnelly, 1981; Flamm, 1981). Reactivity can also be altered when excess oxygen is added to halocarbon plasmas and competition between volatile halide and involatile oxide formation ensues: e.g. in CF_4/O_2 etching of Si (Mogab et al., 1978) and in CCl_3F/O_2 etching of GaAs or InP (Burton et al., 1983). One way in which selectivity can be changed was shown by Ephrath (1979) who added H_2 to a CF_4 discharge and enhanced the SiO_2 : Si etching-rate ratio. Presumably this enhancement results from a decrease in the concentration of free fluorine, which is probably the primary etchant for Si, by formation of HF (Truesdale et al., 1980; Truesdale and Smolinsky, 1979) and from an increase in the concentration of fluorocarbon radicals, which are probably the primary etchants for SiO_2 (Heinecke, 1975). Finally, halocarbons can be used to passivate surfaces, so that anisotropic ion transport can be utilized to effect profile control. Mogab and Levinstein (1980) demonstrated this by etching polysilicon with C_2F_6/Cl_2 and CF_3Cl plasmas.

3.2. III–V plasma etching: importance of group III product desorption

The unifying principles summarized above apply equally well to the etching of III–V compounds and other materials such as Si or Al. However, the situation is complicated by the diverse chemistries of the group III and group V elements; this diversity can affect reactant adsorption, product formation, and product desorption. Reactant choice, for example, is dictated by all of these factors. In particular, the requirement for volatile product formation has limited the choice of reagents to those containing chlorine, bromine, or hydrogen (see tables 2 and 3): e.g. group V fluorides are volatile but group III fluorides are relatively involatile. Little is known about reactant adsorption and product formation, although chlorine is known to chemisorb readily on III–V surfaces (Margaritondo et al., 1979; Montgomery et al., 1978); but fortunately, as we shall see below, these steps are relatively unimportant compared to reactant transport, surface passivation (e.g. by native oxides), and product desorption.

3.2.1. Product desorption
Even with a prudent choice of reactants there is a substantial difference in the relative volatilities of group III and group V products (see table 2). For this reason we expect product desorption to be the rate-limiting step

Table 2
Normal boiling points of possible etching products[a].

Product	Boiling point (°C)	Product	Boiling point (°C)
AlF_3	1291	PF_5	− 75
$AlCl_3$	183	PF_3	−101
$AlBr_3$	263	PCl_5	162
		PCl_3	76
GaF_3	1000	PBr_5	106
$GaCl_2$	535	PBr_3	173
$GaCl_3$	201	PH_3	− 88
$GaBr_3$	279		
Ga_2H_6	139	AsF_5	− 53
		AsF_3	− 63
InF_3	> 1200	$AsCl_3$	130
$InCl$	608	$AsBr_3$	221
$InCl_2$	560	AsH_3	− 55
$InCl_3$	600		
$InBr$	662	SbF_5	150
$InBr_2$	632	SbF_3	319
$InBr_3$	sub	$SbCl_3$	283
		$SbCl_5$	79
		$SbBr_3$	280
		SbH_3	− 17

[a]Handbook of Chemistry and Physics (CRC Press, Boca Raton, FL).

Table 3
III–V plasma etching[a].

Substrate	Gases	References
GaAs	CCl_2F_2 (O_2,Ar,He)	Hu and Howard (1980), Smolinsky et al. (1981), Klinger and Greene (1981), Burstell et al. (1977), Hikosaka et al. (1981), Curran and McCulloch (1982)
	CCl_4 (O_2,Cl_2)	Burton and Smolinsky (1982), Gottscho et al. (1982), Smolinsky et al. (1981), Powell (1982), Smolinsky et al. (1983)
	Cl_2	Burton and Smolinsky (1982), Smolinsky et al. (1981), Donnelly et al. (1982a), Ibbotson et al. (1983)
	Cl_2/Ar^+	Lincoln et al. (1982), Geis et al. (1981)
	$Cl_2/Cl^+/Cl_2^+$	Barker et al. (1982)
	CCl_3F (O_2)	Burton et al. (1983)
	$COCl_2$	Smolinsky et al. (1981)

Table 3 (cont'd)

Substrate	Gases	References
GaAs (cont'd)	PCl_3	Smolinsky et al. (1981)
	HCl	Smolinsky et al. (1981)
	BCl_3/Cl_2	D'Asaro et al. (1981)
	Br_2	Ibbotson et al. (1983)
	H_2	Chang and Darack (1981), Chang et al. (1982), Clark and Fok (1981)
GaP	CCl_4 (O_2)	Burton and Smolinsky (1982)
	Cl_2	Burton and Smolinsky (1982)
GaSb	H_2	Chang et al. (1982)
InP	CCl_2F_2 (Ar,O_2)	Hu and Howard (1980), Curran and McCulloch (1982)
	Cl_2 (O_2)	Burton and Smolinsky (1982), Coldren et al. (1980), Coldren et al. (1981), Bosch et al. (1981), Coldren and Rentschler (1981), Coldren et al. (1982a), Coldren et al. (1982b), Donnelly et al. (1982a)
	$CCl_4/O_2, Cl_2$	Burton et al. (1980), Burton and Smolinsky (1982), Gottscho et al. (1982), Smolinsky et al. (1983)
	$Cl_2/Cl^+/Cl_2^+$	Barker et al. (1982)
	HCl	Smolinsky et al. (1983)
	CCl_3F (O_2)	Burton et al. (1983)
	H_2	Chang et al. (1982), Tu et al. (1982), Clark and Fok (1981)
InGaAsP	Cl_2 (O_2,Ar)	Coldren et al. (1980), Coldren et al. (1981), Coldren et al. (1982a), Coldren et al. (1982b)
	CCl_4/O_2	Burton et al. (1980)
GaAlAs	CCl_2F_2/He	Burstell et al. (1977), Hikosaka et al. (1981)
GaAs oxide	CCl_4	Smolinsky et al. (1981)
	CCl_2F_2	Smolinsky et al. (1981)
	PCl_3	Smolinsky et al. (1981)
	HCl	Smolinsky et al. (1981)
	H_2	Chang and Darack (1981), Chang et al. (1982)

[a]See also tables 1, 2, in Donnelly et al. (1983).

when ample reactant flux is provided and etch-inhibiting deposits are absent. Barker et al. (1982) showed this by using a reactive ion beam comprised of Cl, Cl_2, Cl^+, and Cl_2^+ to study the etching of GaAs and InP as a function of ion energy and flux. From *in situ* Auger electron spectroscopy,

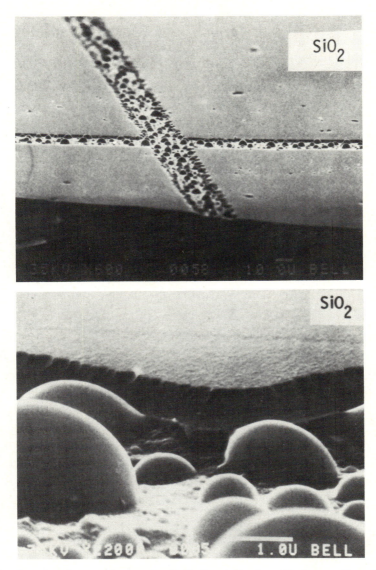

Fig. 8. Scanning electron micrographs of InP after etching in an H_2 plasma with a substrate temperature $< 50°C$ (from Chang et al., 1982).

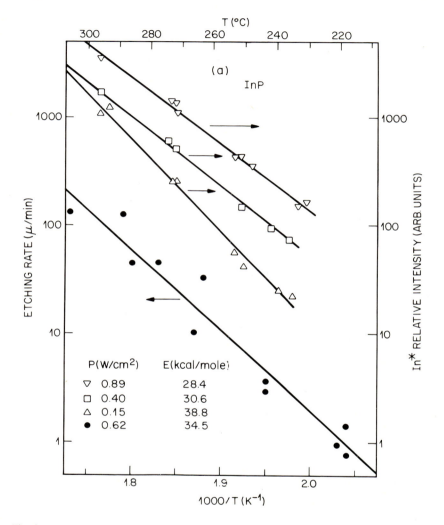

Fig. 9. Arrhenius plot of etching rate (closed symbols) and emission from group III atoms (open symbols, proportional to etching rate) of (a) InP and (b) (facing page) GaAs in a 0.3 Torr Cl_2 discharge. Applied r.f. frequencies were 250 kHz for InP (a) and 13 MHz for GaAs (b). Curve (a) from Donnelly et al. (1982a). Curve (b) courtesy of V.M. Donnelly and D.L. Flamm, AT&T Bell Laboratories.

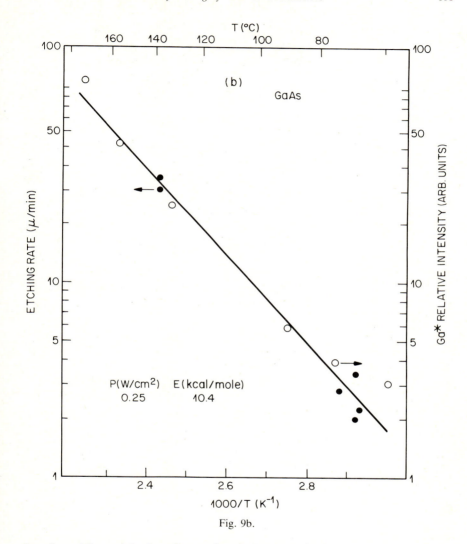

Fig. 9b.

they found In enriched surfaces when InP was etched at room temperature with a 500 eV beam at 30° angle of incidence and an ambient pressure of ~10^{-4} Torr. GaAs and GaP surfaces, etched under similar conditions, remained stoichiometric. At low background chlorine pressures (7 × 10^{-5} Torr) and high ion energies (500 eV), the GaAs and InP etching rates were comparable (~2200Å/min at a current density of 1 mA/cm^2) because in both cases the rates were limited by reactant supply. At low ion energies (200 eV) and high background chlorine pressures (2×10^{-4} Torr), where ion-assisted desorption is less important, the etching rate for GaAs was twice that for InP because of the greater volatility of the Ga chlorides over their In counterparts.

Chang and coworkers (Tu et al., 1982; Chang et al., 1982; Chang and Darack, 1981) etched GaAs, GaSb, and InP in an inductively coupled, r.f. (30 MHz), H$_2$ plasma at pressures between 0.1 and 0.3 Torr. GaAs surfaces etched at temperatures below 150°C showed no signs of preferential As loss. Similar results were obtained for GaSb. However, InP surfaces etched at temperatures below 50°C (and particularly at higher power) showed substantial segregation and In enrichment in the form of μm-sized globules (see fig. 8). Clark and Fok (1981) also produced an overlayer of In metal when InP was etched in an H$_2$ plasma. Both of these experiments were performed at high frequency (> 10 MHz) and low temperature where ion and thermal-induced desorption are minimal. Operation of the H$_2$ discharge at lower frequency (to increase ion energies) and higher temperature (to increase product volatility) may reduce the problem of In segregation associated with etching InP.

Donnelly and coworkers (Donnelly et al., 1982a; Ibbotson et al., 1982) found In and Ga enriched surfaces when InP and GaAs, respectively, were etched in Cl$_2$ plasmas at 0.3 Torr over a substrate temperature range from 200 to 300°C. For all temperatures, Ga enrichment and submonolayer Cl coverage were evident from post-etch Auger analysis of the GaAs surfaces. On the other hand, InP surfaces showed multilayer (~30 Å) InCl$_3$ coverage after processing at low temperatures (210°C); but relatively little adsorbed chlorine was evident for substrates etched at higher temperatures ($\gtrsim$ 250°C). In any case, the InCl$_3$ layers were removed by rinsing with deionized water.

Donnelly et al. (1982a) also found that etching rates as a function of temperature exhibited Arrhenius behavior with activation energies nearly equal to the group III trichloride heats of sublimation (see fig. 9). The InP etching rates were also commensurate with calculated InCl$_3$ rates of evaporation, suggesting that InCl$_3$ product desorption is rate-limiting. However, GaAs etching rates were orders of magnitude smaller than calculated GaCl$_3$ evaporation rates and Donnelly et al. (1982a) proposed that a surface reaction, perhaps dimerization to form Ga$_2$Cl$_6$ was rate-limiting in this case.

3.2.2. Ion-enhanced etching and profile control

So far, we have seen that the disparate volatilities of the group III and group V products often lead to non-stoichiometric etching. In particular, group III product desorption is frequently rate-limiting. Now we consider the ion-neutral synergism and how it influences product desorption which in turn influences profile control.*

Ion-enhanced etching of III–V materials has been demonstrated by Donnelly and coworkers (Donnelly et al., 1981, 1982a, b; Ibbotson et al., 1982) who varied r.f. frequency and power while etching GaAs and InP in Cl_2 plasmas, by Hu and Howard (1980) and Coldren and Rentschler (1981) who varied pressure in the reactive ion etching of GaAs and InP with $CCl_2F_2/O_2/Ar$ and Cl_2/O_2, respectively, and by Mayer et al. (1982) who varied ion energy while etching GaAs and InP in Cl_2 reactive ion beams. All of these experiments altered either the ion transport anisotropy, ion energy, or both.

In the temperature study mentioned in section 3.2.1, Donnelly et al. (1982a) found that the activation energy for etching InP, determined by $InCl_3$ product desorption, decreased from 38.8 to 28.4 kcal/mole when the power density (and indirectly ion energy) was increased from 0.2 to 0.9 W/cm^2 (see fig. 9). Although the power density can affect the surface temperature and therefore thermal desorption rates, this result suggests that ion-induced product desorption can enhance etching rates.

Donnelly et al. (1981) also found that InP etching rates were sensitive to frequency: above 1 MHz, InP did not etch appreciably in a 0.3 Torr Cl_2 plasma. On the other hand, GaAs etching rates were relatively insensitive to frequency between 250 kHz and 13 MHz (Donnelly et al., 1981; Ibbotson et al., 1982). These frequency effects result from changes in sheath potential and ion energy: at frequencies below ~1 MHz, ions traverse the sheath in a time which is short compared to a quarter period of the applied potential, V_a, and can be accelerated to V_a; above ~1 MHz, ions cannot follow the applied field and strike the surface with energies less than the average sheath potential (V_a/π) (Bruce, 1981). Furthermore, Donnelly et al. (1982b) found the applied voltage (and by inference, the ion energy)

---

* Vertical walls have been obtained for GaAs by etching in Br_2 or Cl_2 plasmas at pressures between 0.15 and 0.3 Torr and power densities below 0.5 W/cm^2 (Ibbotson et al., 1983). However, the vertical profiles formed are a result of crystallographic differences in chemical reactivity and are apparently not associated with ion-enhanced desorption. The total amount of undercutting is significant and is dependent on crystallographic orientation.

at constant power to increase four-fold as the frequency decreased from 13 MHz to 250 kHz. Thus, these results are consistent with an increased importance of ion-assisted desorption in removing the less volatile $InCl_x$ products relative to the more volatile $GaCl_x$ products. An alternative explanation for the inability to etch InP at high frequency and high pressure could be inhibition by native oxide.

At lower pressures (10^{-3} to 10^{-2} Torr), in the so-called reactive ion etching (RIE) regime, anisotropic etching of both GaAs and InP was observed with $CCl_2F_2/Ar/O_2$ (Hu and Howard, 1980) and Cl_2/O_2 (Bosch et al., 1981; Coldren and Rentschler, 1981) plasmas. However, the profiles obtained were often positively sloped or overcut (see fig. 7, above). Similar results were obtained at even lower pressures (6×10^{-5} Torr) while reactive ion beam etching GaAs with CCl_4 (Powell, 1982). To date, no satisfactory explanation for the overcut profile has been offered: it is not dependent upon crystallographic orientation nor can it be explained solely by a physical sputtering mechanism (Coldren and Rentschler, 1981; Powell, 1982); Charging phenomena have been suggested as an explanation (Coldren and Rentschler, 1981), although in a recent experiment designed to prevent mask charging an overcut was still observed (Powell, 1982). The anisotropy is sensitive to pressure (Hu and Howard, 1980; Coldren and Rentschler, 1981), feedstock composition (Coldren and Rentschler, 1981), and angle of incidence with respect to ion bombardment (Bosch et al., 1981; Coldren and Rentschler, 1981; Powell, 1982). At low pressures (10^{-3} to 10^{-2} Torr), Hu and Howard (1980) were able to minimize the overcut and achieve nearly vertical walls. They noted a greater dependence of the wall angle on pressure for InP than for GaAs, which is consistent with the $InCl_x$ products being less volatile than their Ga counterparts and InP etching being more sensitive to ion-assisted desorption. Coldren and Rentschler (1981) found a similar dependence of wall angle on pressure and also report that more nearly vertical walls are obtained at elevated temperatures and powers, although this is accompanied by increased surface roughness.

Finally, direct evidence for ion-enhanced etching was obtained by Barker et al. (1982) who found that the InP and GaAs etching rates (normalized to current density) in a Cl_2 reactive ion beam increased by factors of 10 and 5, respectively, when ion energy was increased from 200 to 500 eV and neutral reactant flux to the surface was kept constant. The larger enhancement for InP relative to GaAs is again indicative of the reduced volatility of the In products.

3.2.3. Surface roughening and native oxide removal

Roughened surfaces have been reported in almost all of the III–V etching studies to date (table 3, above). Some of the causes include: incomplete or non-uniform product desorption, variations in the material stoichiometry (e.g. inclusions), non-uniform removal of native oxide, preferential etching at surface defects, ion-induced damage, mask irregularities, redeposition of etched products and masking by non-volatile contaminants. It has been difficult to distinguish between these possibilities, and no systematic investigation has been pursued.

Changes in surface roughness with substrate temperature have been reported in two instances (Coldren and Rentschler, 1981; Donnelly et al., 1982a). In the plasma etching regime, InP and GaAs surfaces are reported to change from rough to smooth as the substrate temperature exceeds 250°C and 120°C, respectively (Donnelly et al., 1982a). This result suggests that group III chloride products build up and non-uniformly mask the underlying substrate. A more extreme case is the H_2 plasma etching of InP at low temperatures ($< 50°C$) where large globules of In metal were observed (see fig. 8, above) (Chang et al., 1982).

Native oxide can inhibit the etching of III–V compounds (Burton and Smolinsky, 1982; Smolinsky et al., 1981) which in turn can result in surface roughening. Smolinsky et al. (1981) report that at 13 MHz and 0.2 to 0.5 Torr, Cl_2 plasmas do not etch GaAs oxide, but chlorocarbon (except for $COCl_2$) plasmas do etch the oxide. Plasmas of HCl also etch GaAs oxide with little surface roughening. However, in every case the oxide is etched more slowly than the underlying substrate. Etch inhibition and associated surface roughness were attributed to non-uniform oxide removal when InP or GaAs were etched in CCl_4 plasmas at low frequency (Burton and Smolinsky, 1982); the non-uniformity could result either from thickness variation in the native oxide or from segregation of the group III and group V elements. Chang et al. (1982) found that substantial In segregation occurred when InP oxide was etched in an H_2 plasma but segregation was not evident when GaAs oxide was etched. Again, this is consistent with the likely lower volatility of the In products. Etching rates in Cl_2 plasmas vary by factors of 2 or 3, probably because chlorine does not etch the oxides efficiently (Gottscho et al., 1982; Donnelly et al., 1982a). By contrast, CCl_4 etching rates, under comparable conditions, are reproducible to better than 10%; this is consistent with similar studies of CCl_4/Al etching (Hess, 1981; Tokunaga et al., 1981; Tokunaga and Hess, 1980; Schwartz and Schaible, 1980), and the probable reduction of group III oxide by chlorocarbon radicals (Gottscho et al., 1982).

3.2.4. Reactants

It is apparent that group III chlorides and bromides are sufficiently volatile that Cl and Br atoms can etch III-V materials. But what happens in more complicated, halocarbon plasmas? Are there other products with sufficient volatility that can be formed from halocarbon radicals?

In CCl_4/O_2 plasmas, GaAs and GaP etching rates were found to correlate with Cl atom emission intensity regardless of the O_2 concentration; and, in fact, the correlation obtained was identical to that for Cl_2/O_2 plasmas under otherwise identical conditions (see fig. 10).* These results

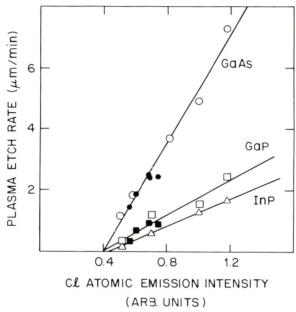

Fig. 10. Plasma etching rates of GaAs (circles), GaP (squares), and InP (triangles) plotted as a function of Cl atomic emission intensity observed in 0.05 Torr, 0.23 W/cm^2, 55 kHz, CCl_4/O_2 (closed symbols) and Cl_2/O_2 (open symbols) discharges. Note that for InP only data from the latter discharge are presented. (From Burton and Smolinsky, 1982).

---

* Although emission intensities cannot be used in general to determine atomic concentrations as a function of feedstock composition, spatially resolved emission experiments suggest that in CCl_4/O_2 mixtures the electron energy distribution does not change appreciably for O_2 concentrations below 50% (Gottscho et al., 1983); and, therefore, Cl emission can be used as a qualitative measure of Cl concentration as a function of CCl_4/O_2 flow ratio in the feedstock.

suggest that Cl atoms are the primary etchants for GaAs and GaP in chlorocarbon discharges. However, InP etching rates did not correlate with Cl atom emission intensity in CCl_4/O_2 plasmas but were higher than expected, suggesting that another etchant or etching mechanism was at play (Burton and Smolinsky, 1982). Further evidence was provided in a subsequent study (Gottscho et al., 1982) where the CCl_4 discharge was operated under conditions which generated copious CCl_x deposits. The deposited layer inhibited etching in general but had a smaller inhibitive effect on InP than on GaAs. It was suggested that formation of perchloro-organometallic compounds such as $InCl_2$ (CCl_3) could enhance etching. To date, the importance of these compounds as products in III–V etching remains an unanswered question.

Summary

The mechanistic investigations of III–V plasma and reactive ion etching discussed above show that many of the explanations used to describe Si and Al etching also apply here. In particular, the ion-neutral synergism and gas-additive effects are similar to those observed previously. However, there is one characteristic that distinguishes III–V from Si and Al: their binary nature. Differences in the relative reactivity and volatility of products for the group III and V elements can lead to non-stoichiometric surface features, which in turn can have catastrophic impact on the utility of the III–V plasma etching process. Minimizing these differences has been the governing factor in the choice of reactants and plasma operating conditions. Reference to table 2, section 3.2, illustrates this point for the case of reactant choice. Of the group III halides, only the chlorides and bromides are reasonably volatile. The fluorides are so involatile that fluorine containing gases are the reactant of choice for selectively etching layers of Si_3N_4 or SiO_2 on III–V compounds. The iodides (not listed) are in general less volatile due to their increased molecular weight and present problems in gas handling. For etching Ga containing compounds, another possible reaction choice is hydride formation. However with In containing compounds, experience has shown that substantial In segregation will result.

Even with a judicious choice of reactants, the relative differences in the reactivity and in the volatility of products for the group III and V elements can be striking. In an effort to circumvent these differences, two basic approaches have been taken in selecting plasma operating conditions: (1) the use of elevated substrate temperatures to increase reactivity and product vapor pressures to the point where some other step (e.g. reactant flux) will become rate limiting; and (2) the use of energetic ion bombardment to stimulate reaction and desorption of products to the same point. While these approaches have been successful in making the III–V plasma etching process viable, each of them introduces new problems. For example, increased ion bombardment in combination with sample motion can be used to achieve anisotropic etching, relatively smooth surface morphology, and efficient removal of native oxides; however, it also entails slow etching, introduces a good deal of surface damage, and may require the use of metal masks which can be undesirable from a processing standpoint. Elevated temperatures, on the other hand, can be used instead of ion bombardment in order to minimize surface damage and achieve rapid etching; but the advantages associated with increased ion bombardment are sacrificed and the high temperatures themselves can be detrimental to some device structures.

Clearly, no single solution for all of the problems associated with III–V plasma etching has surfaced yet. Movement toward such a solution will require a better understanding of the problems. If the causes of surface roughening can be isolated, then perhaps a means other than multidirectional high-energy ion bombardment can be found for eliminating it. If reactants (e.g. halocarbon radicals) can be identified which reduce relative reactivity differences between the group III and group V elements or the semiconductor and its oxide, then improvements might be made on a broad front. If the actual etching products and reaction mechanism on the surface can be identified, then chemical as well as physical means might be developed for stimulating product removal. If the causes of negative undercut can be understood, then changes in the mask or operating conditions might eliminate it. The key to the future utility of III–V plasma etching lies in our ability to answer these questions.

In conclusion, the plasma etching of III–V has had only a minor impact on III–V device processing. In an examination of eight potential applications, plasma etching proved useful in three. If it is to assume a more prominent role, a variety of problems must be solved.

Acknowledgements

The authors gratefully acknowledge many helpful discussions with R.P.H. Chang, L.A. Coldren, R.S. Freund, P.A. Kohl, O.G. Lorimor, L. Marchut, R.H. Saul, and C.B. Zarowin. We also wish to thank C.B. Zarowin for providing results prior to publication.

References

Barker, R.A., T.M. Mayer and R.H. Burton, 1982, Surface Composition and Etching of III–V Semiconductors in Cl_2 Ion Beams, Appl. Phys. Lett. **40**, 583.

Bar-Lev, A., 1979, Semiconductors and Electronic Devices (Prentice/Hall, London).

Bergh, A.A., and P.J. Dean, 1976, Light-Emitting Diodes (Clarendon Press, Oxford).

Bosch, M.A., L.A. Coldren and E. Good, 1981, Reactive Ion Beam Etching of InP with Cl_2, Appl. Phys. Lett. **38**, 264.

Bruce, R.H., 1981, Ion Response to Plasma Excitation Frequency, J. Appl. Phys. **52**, 7064.

Burrus, C.A., and B.I. Miller, 1971, Small-Area, Double-Heterostructure Aluminum–Gallium Arsenide Electroluminescent Diode Sources for Optical-Fiber Transmission Lines, Opt. Commun. **4**, 307.

Burstell, C.B., R.Y. Hung and P.G. McMullin, 1977, Preferential Etch Scheme for GaAs–GaAlAs, IBM Tech. Disclosure Bull. **20**, 2451.

Burton, R.H., and G. Smolinsky, 1982, CCl_4 and Cl_2 Plasma Etching of III–V Semiconductors and the Role of Added O_2, J. Electrochem. Soc. **129**, 1599.

Burton, R.H., H. Temkin and V.G. Keramidas, 1980, Plasma Separation of InGaAsP/InP Light-Emitting Diodes, Appl. Phys. Lett. **37**, 411.

Burton, R.H., C.L. Hollien, L. Marchut, S.M. Abys, G. Smolinsky and R.A. Gottscho, 1983, Etching of Gallium Arsenide and Indium Phosphide in RF Discharges Through Mixtures of Trichlorofluoromethane and Oxygen, J. Appl. Phys. **54**, 6663.

Chang, R.P.H., and S. Darack, 1981, Hydrogen Plasma Etching of GaAs Oxide, Appl. Phys. Lett. **38**, 898.

Chang, R.P.H., C.C. Chang and S. Darack, 1982, Hydrogen Plasma Etching of Semiconductors and their Oxides, J. Vac. Sci. Technol. **20**, 45.

Clark, D.T., and T. Fok, 1981, Surface Modification of InP by Plasma Techniques Using Hydrogen and Oxygen, Thin Solid Films **78**, 271.

Coburn, J.W., 1982, Plasma-Assisted Etching, Plasma Chem. Plasma Process. **2**, 1.

Coburn, J.W., and H.F. Winters, 1979a, Plasma Etching – A Discussion of Mechanisms, J. Vac. Sci. Technol. **16**, 391.

Coburn, J.W., and H.F. Winters, 1979b, Ion and Electron Assisted Gas Surface Chemistry – An Important Effect in Plasma Etching, J. Appl. Phys. **50**, 3189.

Coldren, L.A., and J.A. Rentschler, 1981, Directional Reactive-Ion-Etching of InP with Cl_2 Containing Gases, J. Vac. Sci. Technol. **19**, 225.

Coldren, L.A., K. Iga, B.I. Miller and J.A. Rentschler, 1980, GaInAsP/InP Stripe-Geometry Laser with a Reactive-Ion Etched Facet, Appl. Phys. Lett. **37**, 681.

Coldren, L.A., B.I. Miller, K. Iga and J.A. Rentschler, 1981, Monolithic Two-Section GaInAsP/InP Active-Optical-Resonator Devices Formed by Reactive Ion Etching, Appl. Phys. Lett. **38**, 315.

Coldren, L.A., K. Furuya, B.I. Miller and J.A. Rentschler, 1982a, Etched Mirror and Groove-Coupled GaInAsP/InP Laser Devices for Integrated Optics, IEEE J. Quantum Electron. **QE-18**, 1679.

Coldren, L.A., K. Furuya, B.I. Miller and J.A. Rentschler, 1982b, Combined Dry and Wet Etching Techniques to Form Planar (011) Facets in GaInAsP/InP Double Heterostructures, Electron. Lett. **18**, 235.

Curran, J.E., and D.J. McCulloch, 1982, Profiles Formed by Reactive Ion Etching, in:

Plasma Processing, eds. J. Dieleman and R.G. Frieser (The Electrochemical Society, Pennington, NJ) p. 261.

D'Asaro, L.A., A.D. Butherus, J.V. DiLorenzo, D.E. Iglesias and S.H. Wemple, 1981, Plasma-Etched Via Connections to GaAs FET's, in: Proc. Symp. GaAs and Related Compounds (American Inst. Phys. Conf. Ser. No. 56) p. 267.

Donnelly, V.M., and D.L. Flamm, 1981, Anisotropic Etching in Chlorine-Containing Plasmas, Solid State Technol. April, 1981, p. 161.

Donnelly, V.M., D.L. Flamm and G.J. Collins, 1981, Studies of Plasma Etching of III–V Compounds, Using In Situ Optical Diagnostic Techniques, Electrochem. Soc. Fall Meeting, Denver, Colorado, Extended Abstract October, p. 621.

Donnelly, V.M., D.L. Flamm, C.W. Tu and D.E. Ibbotson, 1982a, Temperature Dependence of InP and GaAs Etching in a Chlorine Plasma, J. Electrochem. Soc. **129**, 2535.

Donnelly, V.M., D.L. Flamm and G. Collins, 1982b, Laser Diagnostics of Plasma Etching: Measurement of Cl_2^+ in a Chlorine Discharge, J. Vac. Sci. Technol. **21**, 817.

Donnelly, V.M., D.L. Flamm and D.E. Ibbotson, 1983, Plasma Etching of III–V Compound Semiconductors, J. Vac. Sci. Technol. **A1**, 626.

Ephrath, L.M., 1979, Selective Etching of Silicon Dioxide Using Reactive Ion Etching with Fluoromethane-Molecular Hydrogen, J. Electrochem. Soc. **126**, 1419.

Flamm, D.L., 1981, Mechanisms of Radical Production in CF_3Cl, CF_3Br and Related Plasma Etching Gases: The Role of Added Oxidants, Plasma Chem. Plasma Process. **1**, 37.

Flamm, D.L., and V.M. Donnelly, 1981, The Design of Plasma Etchants, Plasma Chem. Plasma Process. **1**, 317.

Geis, M.W., G.A. Lincoln, N. Efremow and W.J. Piacentini, 1981, A Novel Anisotropic Dry Etching Technique, J. Vac. Sci. Technol. **19**, 1390.

Gottscho, R.A., G. Smolinsky and R.H. Burton, 1982, Carbon Tetrachloride Plasma Etching of GaAs and InP: A Kinetic Study Utilizing Nonperturbative Optical Techniques, J. Appl. Phys. **53**, 5908.

Gottscho, R.A., G.P. Davis and R.H. Burton, 1983, Spatially-Resolved Laser-Induced Fluorescence and Optical Emission Spectroscopy of Carbon Tetrachloride Glow Discharges, Plasma Chem. Plasma Process. **3**, 193.

Heinecke, R.A.H., 1975, Control of Relative Etch Rates of SiO_2 and Si in Plasma Etching, Solid State Electron. **18**, 1146.

Hess, D.W., 1981, Solid State Technol., April, 1981, p. 189.

Hikosaka, K., T. Mimura and K. Joshin, 1981, Selective Dry Etching of AlGaAs–GaAs Heterojunction, Japan. J. Appl. Phys. **20**, L847.

Hindin, H.J., and J.G. Posa, 1982, Gallium Arsenide Forges Ahead Toward Very Large-Scale Integration, Electronics, Feb. 24, 1982, p. 111.

Hu, E.L., and R.E. Howard, 1980, Reactive-Ion Etching of GaAs and InP Using CCl_2F_2/Ar/O_2, Appl. Phys. Lett. **37**, 1022.

Ibbotson, D.E., D.L. Flamm and V.M. Donnelly, 1983, Crystallographic Etching of GaAs with Bromine and Chlorine Plasmas, J. Appl. Phys. **54**, 5974.

Ibbotson, D.E., D.L. Flamm, V.M. Donnelly and B.S. Duncan, 1982, Summary Abstract: Studies of Plasma Etching of III–V Compounds: The effects of Temperature and Discharge Frequency, J. Vac. Sci. Technol. **20**, 489.

Iga, K., and B.I. Miller, 1982, Chemically Etched-Mirror GaInAsP/InP Lasers – Review, IEEE J. Quantum Electron. **QE-18**, 22.

Keramidas, V.G., G.W. Berkstresser and C.L. Zipfel, 1980, Planar, Fast, Reliable Single-Heterojunction Light-Emitting Diodes for Optical Links, Bell Syst. Tech. J. **59**, 1549.

Klinger, R.E., and J.E.. Greene, 1981, Reactive Ion Etching of GaAs in CCl_2F_2, Appl. Phys. Lett. **38**, 620.

Kohl, P.A., C. Wolowodiuk and F.W. Ostermayer, 1983, The Photoelectrochemical Oxidation of (100), (111) n-InP and n-GaAs, J. Electrochem. Soc. **130**, 2288.

Lincoln, G.A., M.W. Geis, L.J. Mahoney, A. Chu, B. Vojak, K.B. Nichols, W.J. Piacentini, N. Efremow and W.T. Lindley, 1982, Ion Beam Assisted Etching for GaAs Device Applications, J. Vac. Sci. Technol. **20**, 786.

Mandal, R.P., 1982, III–V Semiconductor Integrated Circuits – A Perspective, Solid State Technol., Jan., 1982, p. 94.

Margaritondo, G., J.E. Rowe, C.M. Bertoni, C. Calandra and F. Manghi, 1979, Chemisorption Geometry on Cleaved III–V Surfaces: Cl on GaAs, GaSb and InSb, Phys. Rev. **B20**, 1538.

Mogab, C.J., and H.J. Levinstein, 1980, Anisotropic Plasma Etching of Polysilicon, J. Vac. Sci. Technol. **17**, 721.

Mogab, C.J., A.C. Adams and D.L. Flamm, 1978, Plasma Etching of Si and SiO_2 – The Effect of Oxygen Additions to CF_4 Plasmas, J. Appl. Phys. **49**, 3796.

Montgomery, V., R.H. Williams and R.R. Varma, 1978, The Interaction of Chlorine with Indium Phosphide Surfaces, J. Phys. **C11**, 1989.

Mucha, J.A., and D.W. Hess, 1983, Plasma Etching, in: Microcircuit Processing: Lithography and Dry Etching, eds. L.F. Thompson and C.G. Willson (American Chemical Society).

Ostermayer, F.W., P.A. Kohl, R.H. Burton and C.L. Zipfel, 1982, Photoelectrochemical formation of Integral Lenses on InP/InGaAsP LED's, Abstracts of Papers, IEEE specialist conference on Light Emitting Diodes and Photodetectors, Ottawa, Canada, Sept. 15–16.

Podlesnik, D., R.M. Osgood, V. Daneu and A. Sanchez, 1982, High Resolution Etching of GaAs and CdS Crystals, Materials Research Society, Annual Meeting, Boston, Mass, Nov. 1–4.

Powell, R.A., 1982, Reactive Ion Beam Etching of GaAs in CCl_4, Japan. J. Appl. Phys., **21**, L 170.

Schwartz, G.C., and P.M. Schaible, 1980, Reactive Ion Etching in Chlorinated Plasmas, Solid State Technol., Nov., 1980, p. 85.

Schwartz, G.C., and P.M. Schaible, 1981, Reactive Ion Etching of Silicon: Temperature Effects, in: *Plasma Processing, eds. R.G. Frieser and C.J. Mogab* (The Electrochemical Society, Pennington, NJ), p. 133.

Smolinsky, G., R.P. Chang and T.M. Mayer, 1981, Plasma Etching of III–V Compound Semiconductors and Their Oxides, J. Vac. Sci. Technol. **18**, 12.

Smolinsky, G., R.A. Gottscho and S.M. Abys, 1983, Time Dependent Etching of GaAs and InP with CCl_4 on HCl Plasmas: Electrode Material and Oxidant Addition Effects, J. Appl. Phys. **54**, 3518.

Sze, S.M., 1981, Physics of Semiconductor Devices, 2dn Ed. (Wiley, New York).

Tokunaga, K., and D.W. Hess, 1980, Aluminum Etching in Carbon Tetrachloride Plasmas. J. Electrochem. Soc. **127**, 928.

Tokunaga, K., F.C. Redeker, D.A. Danner and D.W. Hess, 1981, Comparison of Aluminum Etch Rates in Carbon Tetrachloride and Boron Trichloride Plasmas, J. Electrochem. Soc. **128**, 851.

Truesdale, E.A., and G. Smolinsky, 1979, The Effect of Added Hydrogen on the RF Discharge Chemistry of CF_4, CF_3H, and C_2F_6, J. Appl. Phys. **50**, 6594.

Truesdale, E.A., G. Smolinsky, and T.M. Mayer, 1980, The Effect of Added Acetylene on the RF Discharge Chemistry of C_2F_6. A Mechanistic Model for Fluorocarbon Plasmas, J. Appl. Phys. **51**, 2909.

Tu, C.W., R.P.H. Chang and A.R. Schlier, 1982, Surface Etching Kinetics of Hydrogen Plasma on InP, Appl. Phys. Lett. **41**, 80.

Zarowin, C.B., 1982, American Vacuum Society 29th National Symposium, Baltimore, MD, Nov. 6–19, 1982, Abstract No. TFTA–09.

CHAPTER 4

REACTIVE ION BEAM ETCHING

R.A. POWELL

Varian Associates, Inc.
Corporate Solid State Laboratory
Palo Alto, California 94303, USA

D.F. DOWNEY

Varian Associates, Inc.
Extrion Division
Gloucester, Massachusetts 01930, USA

Dry Etching for Microelectronics, edited by R.A. Powell
© Elsevier Science Publishers B.V., 1984

Contents

1. Introduction — 115
2. Dry etch technologies — 116
   2.1. Plasma etching — 116
   2.2. Reactive ion etching (RIE) — 118
   2.3. Ion beam milling — 119
3. Reactive Ion Beam Etching (RIBE) — 121
   3.1. General characteristics — 121
   3.2. System design — 126
   3.3. Ion sources — 130
      3.3.1. Broad-beam ion source technology — 130
      3.3.2. Use of reactive gases in a broad-beam ion source — 135
      3.3.3. The effect of a magnetic field — 139
      3.3.4. Use of low energy ions — 142
      3.3.5. Unconventional ion sources — 143
4. Mechanisms of RIBE etching — 147
   4.1. Ion-beam-enhanced gas–surface chemistry — 149
   4.2. Reactive ion bombardment — 155
   4.3. Neutral species — 160
5. Patterning and etching materials with RIBE — 163
   5.1. Aluminum and aluminum alloys — 163
   5.2. Silicon dioxide and silicon — 171
      5.2.1. RIBE with chlorine-containing gases — 171
      5.2.2. RIBE with fluorine-containing gases — 176
      5.2.3. Etching mechanisms for silicon and silicon dioxide — 182
         5.2.3.1. Chlorine-containing gases — 182
         5.2.3.2. Fluorine-containing gases — 183
   5.3. Photoresist and organics — 185
   5.4. New applications — 192
      5.4.1. Refractory metals and metal silicides — 192
      5.4.2. Carbides — 199
      5.4.3. III–V compound semiconductors — 200
6. RIBE-induced damage — 208
7. Conclusion — 210
References — 211

1. Introduction

The development of integrated circuit fabrication in the electronics industry has been accompanied by new and increasingly stringent requirements on etching technology (Tolliver, 1980). In the early years of IC fabrication, wet chemistry had no competition as an etching technique for pattern transfer and selective materials removal. As the feature size of devices decreased, however, the tendency of isotropic chemical etching to undercut masking layers began to limit the resolution of transferred patterns and alternate dry etching techniques were explored.

In the late 1960's, dry etching began to appear in wafer fab lines in the form of photoresist strippers. The introduction of these 'plasma ashing' machines was followed quickly by the development of processes and equipment for plasma etching various other materials, such as silicon dioxide, silicon nitride, single-crystal and polycrystalline silicon. These plasma etching technologies, like wet chemistry before them, had their resolution limited by undercutting, and other, more anisotropic etching techniques appeared. Among these were Reactive Ion Etching (RIE) and ion beam milling with inert gases (Maddox and Splinter, 1980). The RIE technique provided moderate selectivity and etch rates, and, in many cases, high anisotropy. The major disadvantages were lack of control over critical process parameters and reproducibility. The ion beam milling approach provided excellent control over the critical process parameters and yielded high spatial resolution. Unfortunately, the etching was slow and, being nonchemical in nature, rather unselective and prone to physical sputtering artifacts (such as trenching and redeposition).

In the mid to late 1970's, research began on a dry etch technique which would combine positive features of etching with directed ion beams and etching in reactive plasmas. The technology which developed was known as Reactive Ion Beam Etching or RIBE – an acronym pronounced by the authors to rhyme with 'tribe'. The objective of RIBE was to combine the major advantages of the reactive plasma technologies (high selectivity, photoresist integrity, and relatively high etch rates) with those of inert ion beam milling (high resolution, anisotropy, independent control over process parameters) to produce a new dry etching technology capable of addressing the demands of LSI processing and the future, more stringent demands of VLSI processing.

Before describing the RIBE technique, therefore, we briefly review salient features of the dry etch techniques which preceded it and influenced its development.

2. Dry etch technologies

2.1. Plasma etching

In reactive plasma etching, substrates are directly exposed to a chemically-active, partially ionized gas. The composition of the etching gas is chosen to efficiently volatilize the layer to be etched and to provide good selectivity with respect to etch mask and underlying substrate. Usually, the etch gas chosen is chemically nonreactive in its neutral, ground state. Only upon entering the glow discharge region of the plasma where it is decomposed and ionized (into ions, electrons, free radicals, etc.) are reactive species available. The equipment used for this process falls into one of two generic types: the barrel reactor and the parallel plate or planar reactor.

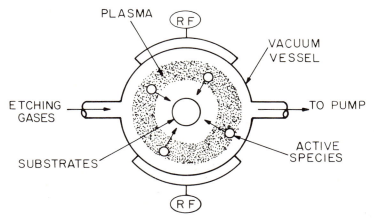

Fig. 1. Capacitively coupled barrel-type RF plasma reactor.

Figs. 14, 15, 38–42, 51–53 reprinted by permission of the publisher, The Electrochemical Society, Inc.; Figs. 8, 32 and 35 were originally presented at the Spring 1983 Meeting of the Electrochemical Society, Inc., held in San Francisco, California.

Figure 1 is an illustration of a *barrel reactor*. Typically, a boat of wafers is placed in a perforated 'Faraday cage' held at ground potential. This cage is centered in a cylindrical vacuum vessel and RF power provided by capacitive (as shown) or inductive coupling. The vacuum vessel is evacuated to a base pressure between 100 mTorr and 1 Torr by a mechanical pump, at which point the appropriate etching gas is fed into the vessel in continuous flow to the vacuum pump. This gas is then decomposed and ionized in the chamber by RF energy. Neutral reactive species are then able to diffuse through the Faraday cage to react with those portions of the substrate not protected by resist. The etching in this case is entirely a chemical process. This technique provides, through the careful selection of reactive gas, excellent selectivity. The major disadvantage of the barrel reactor is that the etching is almost always isotropic. As with any 'pure' chemical process, it is also sensitive to impurities, both in the etch gas and on the wafers prior to etch. A 'de-scum' etch is often required to control surface contamination on the substrates. Other problems of etching with barrel reactors are: etch rate uniformity, temperature control, process reproducibility and end-point detection.

In fig. 2, the parallel-plate, *planar reactor* is represented. It usually consists of two parallel plates, with the power being supplied to the upper plate and the substrates loaded onto a cooled lower plate. The glow discharge is created between the two parallel plates. In this configuration, the plasma is in contact with the substrate, which is thus exposed to low-energy ion and electron bombardment. The advantages of this geometry are: good selectivity, the etching can be anisotropic, and there is better uniformity, reproducibility and temperature control than in the barrel reactor. The disadvantages are: smaller wafer capacity, more radiation damage effects (since the wafers are in contact with the plasma) and, in some cases, the etching may leave residues.

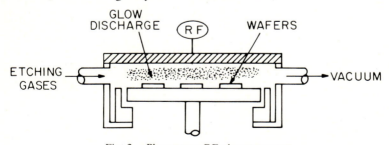

Fig. 2. Planar-type RF plasma reactor.

2.2. Reactive Ion Etching (RIE)

By powering the substrate electrode and lowering the operating pressure of the parallel-plate planar system, another mode for etching can be achieved. This mode is Reactive Ion Etching (RIE). Figure 3 illustrates a typical RIE system. Relatively speaking, these systems have enjoyed great success and have been applied to the etching of most materials of interest

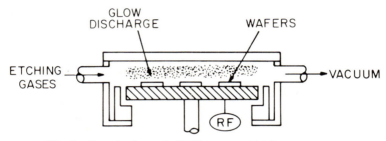

Fig. 3. Reactive ion etch (RIE) system with planar geometry.

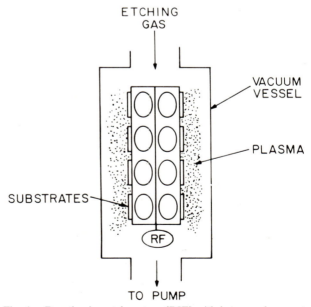

Fig. 4. Reactive ion etch system (RIE) with hexagonal geometry.

in silicon or III–V semiconductor technology. The lower operating pressure in reactive ion etchers, usually in the range of 1×10^{-3} to 1×10^{-1} Torr, lengthens the mean free path of the reactive ions generated in the plasma and permits ions to impinge at near normal incidence upon the substrate, resulting in a considerable improvement in the anisotropy of the etch. In this configuration, however, the etch rate is reduced below that achieved in the higher-pressure reactors because of the reduced density of the etching gas. Other geometrical configurations have been used for RIE besides the early parallel-plate design. Figure 4 illustrates a more recent hexagonal geometry where the RF power is applied to the center electrodes which also support the substrates (Weiss 1982). This geometry has been applied to the etching of many materials of interest in VLSI circuit fabrication.

Problems encountered in RIE include the possibility of radiation damage to sensitive devices and the fact that the etch leaves residues in some important material systems such as AlCuSi. Also, compared to ion beam approaches, there is less control over the critical process parameters (ion energy and current), which affects etch reproducibility. Even with these problems, RIE offers the best overall combination of anisotropy, selectivity and process control of today's plasma-based dry etching technologies.

2.3. Ion beam milling

Another etching technology investigated for semiconductor processing in the early to mid-1970's was ion beam milling. Ion beam milling is a purely physical process in which energetic inert ions (usually 500–1000 eV) are extracted from a broad-beam ion source and delivered to the substrates as a uniform directional beam (see fig. 5). Since etching proceeds via physical sputtering, etch products are nonvolatile and redeposition and trenching are common problems. To alleviate these problems, ion beam milling is typically performed using tilt and rotation of the substrates. This remedy imposes limitations on resolution due to shadowing which results from the off-normal angle of incidence of the beam. Even with these difficulties, ion beam milling has successfully been used to replicate 1000 Å features with high yield. The major advantages of ion beam milling are: high resolution, good uniformity, high anisotropy, residue-free etching, good control over process parameters, and relative freedom from radiation damage effects resulting from the physical isolation of the plasma and the source. The disadvantages are: lack of sensitivity, possible trenching and redepo-

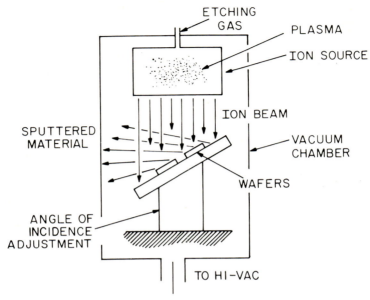

Fig. 5. Inert ion beam milling system. Samples may be tilted and/or rotated with respect to the incident ion beam.

sition, faceting of the photoresist, damage from energetic ion bombardment, and low etch rates resulting in low wafer throughput.

As discussed above, there are inherent difficulties with both plasma etching systems (e.g., control over critical process parameters, residues, etc.) and inert ion milling (e.g., slow etch rates, trenching and redeposition and lack of sensitivity). An attempt to merge these two technologies led to the creation of Reactive Ion Beam Etching. As will be discussed more in later sections, 'RIBE' is a misnomer as the technique is usually performed, since the directed reactive ion beam is only one of several components in the dry etch process. For example, both reactive ion beams and active neutrals (whether produced in the ion source or introduced directly into the etching chamber) are generally present. Depending on the process application, one may wish to vary the relative flux of reactive energetic ions and reactive low-energy neutrals. In this way, the advantages of plasma etching are merged with those of ion beam milling to produce a promising new dry etch technology.

In the present paper, we review the fundamentals of the RIBE technique, its instrumentation, and important applications in microelectronics. For additional reviews of RIBE, the reader is referred to references by Downey and Powell (1983), Bollinger (1983), Stein (1982), Hakhu (1981) and Downey et al. (1981).

3. Reactive Ion Beam Etching (RIBE)

3.1. General characteristics

Table 1 compares the operating characteristics of RIBE with other dry etch techniques. In RIBE, a gas capable of producing a reactive ion beam

Table 1
Comparison of operating characteristics of RIBE with other dry etch techniques.

Characteristic	Dry etch technique				
	Barrel etch	Planar plasma etch	Planar RIE	Ion beam milling	Reactive Ion Beam Etching
Substrate location	Surrounded by plasma	In plasma on grounded electrode	In plasma on driven electrode	In beam, separated from plasma	In beam, separated from plasma
Pressure	0.1–1 Torr	0.1–1 Torr	10-100 mTorr	$\sim 1 \times 10^{-4}$ Torr	$\sim 1 \times 10^{-4}$ Torr
Ion energy (eV)	0	10–100	100-700	300-1500	300-1500
Control of ion energy	None	Semi-independent	Semi-independent	Independent	Independent
Flux	Neutrals	Neutrals; low energy ions	Neutrals; high energy ions	Neutrals; high energy ions	Neutrals; high energy ions
Control of neutral flux	Independent	Semi-independent	Semi-independent	Semi-independent	Semi-independent
Selectivity	Excellent	Very good	Good	Poor	Good
Profile	Isotropic	Often isotropic	Often anisotropic	Anisotropic	Anisotropic
Degree of profile control	None	Low	High	None	Low
Reaction product	Volatile	Volatile	Nonvolatile & volatile	Nonvolatile	Nonvolatile & volatile
Mechanism	Chemical	Chemical	Physical & chemical	Physical	Physical & chemical

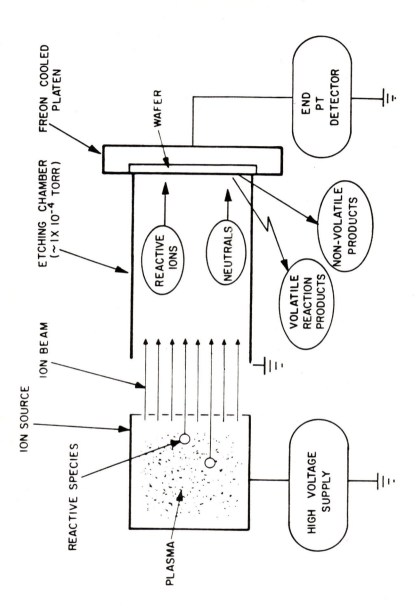

Fig. 6. Representation of RIBE etching with a broad-beam ion source.

is introduced into a broad-beam-type ion source and ionized (refer to fig. 6). The chemical composition of the etchant gas is chosen to efficiently volatilize the layer to be etched and provide good selectivity with respect to the mask and substrate. Reactive ions are then extracted (usually at 200–1000 V) from the source by a pair of independently-controlled grids, and delivered to the substrate as a uniform, collimated reactive ion beam. In addition to these energetic reactive ions (depending on the source gas chosen), neutral reactive species can also diffuse from the source and participate in the etch process. The interaction of low-energy neutral species and energetic reactive ions at the sample surface is discussed in detail in section 4. In general, however, etching is accomplished by a combination of reactive chemistry and physical sputtering at low pressures (usually about 1×10^{-4} Torr). Inherent in the geometry shown in fig. 6 is the capability to independently control the reactive and physical components of the beam. The ion energy and current density can then be independently set to maximize the etch rate, while still providing the appropriate balance between the chemical and physical etching components.

Other variables which can affect the 'ratio' of the chemical to the physical components of the etch process are gas flow, source-to-substrate distance (both of which affect the number of active neutrals reaching the substrate), gas mixture in the ion source, and the residual vacuum environment. Along these lines, the desire to better control ion-beam-assisted chemistry in the etch process has led to the performance of RIBE in two other modes. The first is inert ion-assisted chemical etching (Geis et al., 1981), and the second is reactive ion-assisted chemical etching (Stein, 1982; Chinn et al., 1983a, b).

In inert ion-assisted chemical etching (fig. 7), an inert gas such as argon is introduced directly into a broad-beam ion source and ionized. The ions are then extracted from the source at a given energy by a pair of independently-controlled grids and delivered to the substrate as a spatially uniform, highly directional beam. The chemical component of the etching is provided by a controlled leak of the appropriate gas into the etching chamber. A gas jet close to the sample surface can be used or the etch chamber simply backfilled with the gas. In the former case, very high partial pressures at the sample surface can be produced. The gas chosen is one which can efficiently (in the presence of ion bombardment) volatilize the layer to be etched and provide good selectivity with respect to the mask and substrate. If the background gas chosen is nonreactive in the absence of ion beam bombardment, the etching produced can be very

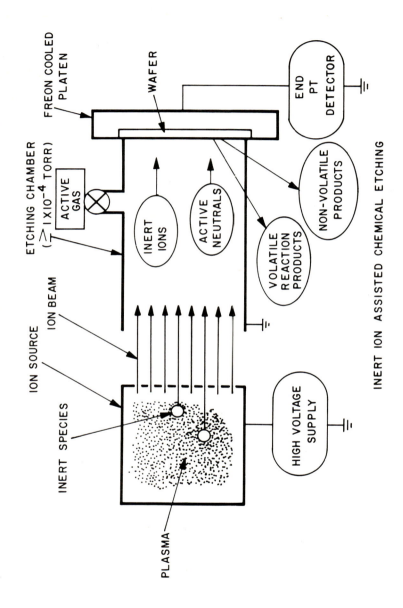

Fig. 7. Ion-assisted chemical etching performed with an inert gas in the ion source and reactive gas background in the etch chamber.

Reactive ion beam etching 125

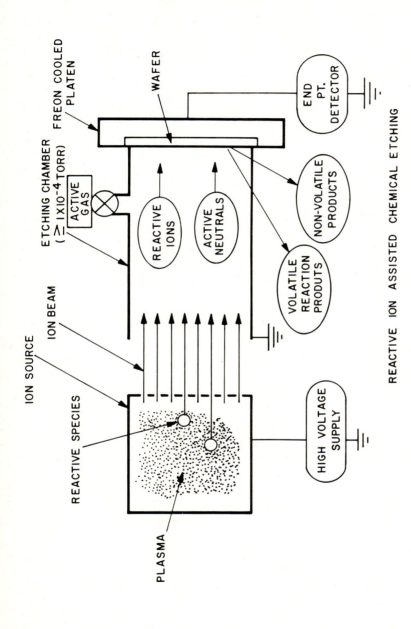

Fig. 8. Ion-assisted chemical etching performed with a reactive gas in the ion source as well as reactive gas background in the etch chamber.

anisotropic. If, however, the gas chosen reacts with the layer to form volatile species without the aid of the ion beam, isotropic etching can be obtained. By exploiting the control one has over the physical component (ion beam current density and energy) and chemical component (flow rate into the etching chamber) with this technique, controlled sidewall profiles have been successfully produced (Geis et al., 1981; Chinn et al., 1983b).

Another variation of RIBE is reactive ion-assisted chemical etching (fig. 8). In this case, a reactive gas such as CCl_4 or XeF_2 is introduced into an ion source, ionized, and extracted as a collimated ion beam, as in conventional RIBE. Additional chemical components to the etching are then introduced independently at the sample surface by use of a directed gas jet or by backfilling the etch chamber, and the etching proceeds in a similar fashion as for inert-ion-assisted chemical etching. In cases where one is limited by the chemical component of the etching, reactive or inert ion-assisted chemical etching provides an advantage over conventionally performed RIBE. In the case where there is a need for a delicate balance between the physical and chemical components, e.g., when etching Al-CuSi, the addition of more chemical components without the correct addition of a physical component can lead to deleterious effects (Downey et al., 1981). In this case, too much chemical etch component can lead to an enhanced Al etch rate without the appropriate increase in the etch rate of Si and Cu. In addition, undercutting of masking layers can become a problem. The choice which mode of RIBE is employed must be considered carefully for each individual case and in light of the desired end result.

3.2. System design

Several designs for reactive ion beam etching systems have been employed over the last several years. In some cases, previously built inert ion beam milling machines have been modified for operation with reactive gases (Bollinger and Fink, 1980; Burggraaf, 1980; Robertson, 1978; Jolly and Clampitt, 1982). In other cases, machine design was developed to accommodate entirely new technology, such as the ECR microwave ion source (Anelva, Model RIB-310) described by Tsukada (1983). In general, however, the major features of commercial RIBE systems with broad-beam ion sources are similar, and a representative RIBE system (Downey et al. 1981) with automatic wafer handling is shown in figs. 9 and 10 (Varian Extrion Model IA-200). It consists of a 15 cm diameter broad-beam ion source which generates a highly uniform ($\pm$ 2–5% over a 4" wafer), colli-

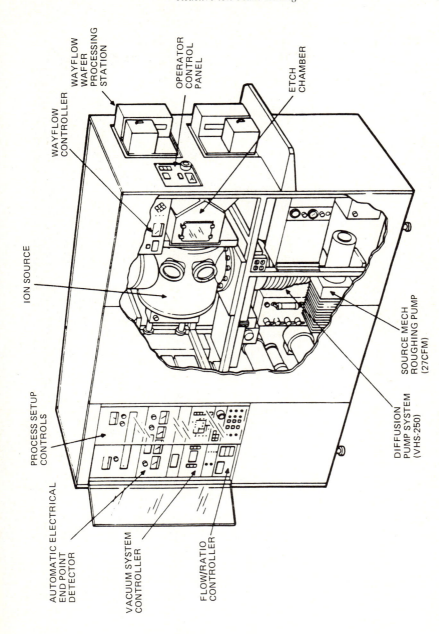

Fig. 9. Illustration of automated commercial RIBE system (Varian/Extrion Model RE-580).

Fig. 10. Detail of the Varian/Extrion Model RE-580 RIBE system shown in fig. 9.

mated reactive ion beam (beam energy ~300–1000 eV); a diffusion-pumped, high-vacuum process chamber backed by a 27 cfm chemical-series rotary pump with a disposable cartridge oil filter; a cryotrap to minimize water vapor partial pressure in the etch chamber; a freon-cooled wafer platen which allows etching at high power density ($\lesssim 1\ \mathrm{W/cm^2}$) without thermal degradation of photoresist; and a digitally controlled, serial process wafer-handling system with vacuum load locks that allows continuous wafer processing without exposing the process chamber to atmospheric contamination. An electrical end-point detector (based on monitored changes in sample secondary electron emission when the underlying stop layer is reached) allows the operator to preset the percent of overetch desired. Modifications of this design to enable etching in the ion-assisted chemical etching mode have been reported (Chinn et al. 1983a, b). This was accomplished by adding a mass-flow-controlled gas inlet onto the etching chamber. With this modification, either mode of etching can readily be carried out.

Operation of the RIBE system shown in fig. 9 is semi-automatic. Process parameters are selected on a simple setup control panel, and the desired overetch is chosen. Cassettes of wafers are placed in the entrance and exit lock carriages and the start button engaged. The wafers are then automatically introduced through the lock–lock mechanism, enter onto the platen and are etched. When the end-point detector senses that the etch is complete, the wafer automatically exits the system and the next enters. When all the wafers are processed, an audio tone signals the operator to load the next cassette.

The system shown in fig. 9 is only one manifestation of the 5 major components comprising a commercial RIBE system, viz: (1) ion source, (2) etch chamber, (3) vacuum pumping, (4) wafer handling (including wafer cooling), and (5) process control (including end-point detection). Ion source design will be treated in detail in section 3.3; however, the usual choice is a broad-beam, Kaufman-type ion source with either a DC or microwave plasma discharge. Since etch chamber pressures for RIBE are in the range of about 10^{-3}–10^{-4} Torr, diffusion or cryopumping can be used. Mechanical backing pumps are built for corrosive gas handling and/or have external recirculating oil filtration systems to prolong pump life. Wafer handling can be batch or serial, although batches of large (4–5″) wafers necessitate very large-diameter ion sources or mechanical rotation of samples through a smaller beam for uniformity. Many systems provide a tilt and rotation of the sample with respect to the incident ion beam, allowing some control over etched sidewall profiles.

Under typical RIBE conditions (500–1000 eV, 0.5–1 mA/cm^2), as much as 1 W/cm^2 can be incident at the wafer surface, leading to thermal degradation of masking layers, unless samples make good thermal contact to a cooled platen. Mechanical clamping (assisted by flow of a thermally conductive gas between the wafer backside and platen), thermally conductive adhesives and electrostatic holddowns to a dry, thermally conductive elastomer have all been employed to improve heat transfer away from the photoresist-coated wafer surface (Egerton et al., 1982).

Finally, we note that the use of end-point detection schemes in RIBE systems has been limited. Since the plasma discharge and sample are widely separated (~15–20 cm), end-point schemes used for other dry etch techniques such as optical emission spectroscopy from the plasma have been difficult to implement. One approach described earlier is to monitor the ratio of the incident current to the electrical current flowing through the sample during RIBE. This ratio is found to be sensitive to the constituent materials at the sample surface and can be used to determine when an underlying stop layer has been reached (Sakaki et al., 1981). This scheme has worked very well with metals such as Al or Al alloys over SiO_2. Another approach is to measure the optical emission from excited species close to the substrate, although this is currently more applicable to research than to production RIBE equipment (Dzioba et al., 1981).

As with other dry etch techniques which have been commercialized (such as RIE and plasma etching), a variety of system configurations is now available to perform RIBE (for some examples, see: Downey et al., 1981; Jolly and Clampitt, 1982; Tsukada, 1983; Anonymous, 1981) and choices between designs must, as always, be made in light of the specific applications planned.

3.3. Ion sources

Although a variety of ion sources has been used to perform RIBE, the most common today are broad-beam ion sources in which a broad (~1–15 cm diameter), collimated ion beam is produced by multiaperture ion optics. These sources are sometimes called 'Kaufman-type' ion sources, in reference to Harold R. Kaufman who pioneered their development in the 1960's. Broad-beam ion sources were originally developed for use as 'ion thrusters' in space propulsion applications. More recently, they have been used for dry etching (both RIBE and inert ion beam milling) and sputter deposition. The reader is referred to articles by Kaufman (1978), Kaufman et al. (1982a, b) and Harper et al. (1982) for in-depth discussions of the technology of such sources and their past and projected applications.

3.3.1. Broad-beam ion source technology
It is difficult to separate the technique of RIBE as it is practiced today from the technology of broad-beam ion sources, since the special characteristics of these ion sources give RIBE many of its unique features. For example, in a Kaufman-type ion source, ions are usually produced by

direct-current electron bombardment of a low-pressure gas ($\sim 10^{-3}$ Torr) rather than by a radio-frequency discharge as is common in plasma or RIE reactors. This means that the composition of the plasma discharge used for RIBE, e.g., the relative populations of reactive ions and free radicals, will not in general be the same as that of the discharge in a plasma or RIE reactor operating at 13.56 MHz, even though the same source gases are used. Also, in a Kaufman-type ion source, unlike plasma and RIE reactors, both ion production and ion acceleration are spatially and electrically isolated from the substrate and walls of the etching chamber. Finally, the broad-beam ion source produces an ion beam whose energy, current density, and direction are, to a large degree, independently controllable. At the ion energies normally used in RIBE, 0.1–1 keV, broad-beam ion sources are capable of supplying ion currents of about 50–150 mA. In this regard, the 15 cm diameter Ion Tech source used in our laboratory (Ion Tech, Inc., Ft. Collins, CO) is representative. Over 100 mA of ion current at 500 eV can be delivered to a 4″ wafer with uniformity of better than ±5% – the wafer being about 15 cm downstream from the source extraction grids. Kaufman-type sources are available with beam diameters ranging from ~ 2.5 cm (Ion Tech Model 2.5-1500-40) to 35 cm (Micro Ion Mill TLA-35; Technics, Ltd., Alexandria, VA). Maximum beam energy is usually 1.5–2 keV, although a source operating at 7.5 keV is commercially available (Technics Model TLA 5.5-II).

A typical broad-beam ion source is shown in fig. 11. The source gas or gas mixture flows into the cylindrical discharge chamber (typical flow rates are approximately 1–10 sccm) where ionization occurs by primary electron bombardment from a thermionic cathode (hot filament). At the low chamber pressures used ($\sim 10^{-3}$ Torr), the mean free path length for electron impact ionization is much larger than the chamber dimensions. A magnetic field is therefore applied so that the primary electrons follow lengthy paths before reaching the anode. In this way, the probability of ionizing collisions with the gas atoms in the source is increased. Although early sources used an axial magnetic field ($\boldsymbol{B}$ field perpendicular to the extraction grids), more recent designs employ a so-called multipole magnetic field configuration. The discharge chamber shown in fig. 11 is such a multipole design. The magnetic field is confined near the outside wall of the discharge chamber and concentrated near the anode in sufficient strength (~ 80 G) to prevent primary electrons from prematurely escaping to the anodes. This configuration results in a $\boldsymbol{B}$ field-free region throughout most of the discharge chamber volume. In this region, both primary elec-

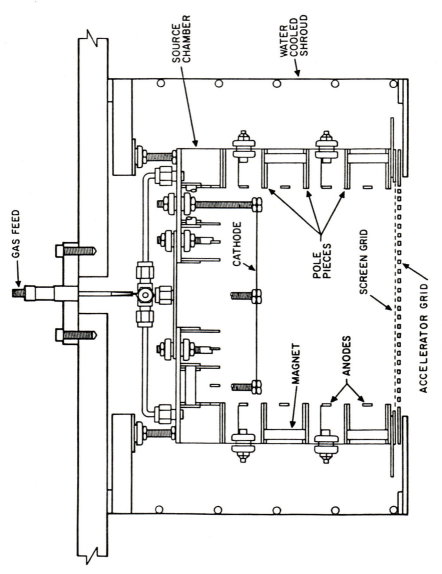

Fig. 11. Broad-beam Kaufman-type ion source.

tron and ion production are uniformly distributed, allowing an ion beam of uniform profile to be readily extracted. Once a discharge is established, the source chamber is filled with a conducting plasma composed of nearly equal numbers of electrons and positive ions in a background of neutral atoms. The electrons, due to their large mobility, preferentially escape to any electrode surface that is not negative with respect to the plasma, resulting in the plasma becoming self-biased slightly positive ($\sim +5$ V) above the most positive chamber surface, i.e., the anode. The ions originate at the plasma potential which, within a few volts, is also the anode potential. Since the plasma potential in this source is close to the anode potential, the primary electrons have energies close to the arc discharge voltage (cathode-to-anode voltage), which is usually about 40–80 eV.

About 10–30% of the ions produced in the source reach the accelerator system and make up the extracted ion beam. The ion beam current is roughly proportional to both the ion production rate and the plasma density in the source. This allows the extracted ion-beam current to be varied independently of ion energy (assuming one has not reached the space-charge-limited current density of the source).

For ion extraction, a two-grid system involving a screen grid (up-stream) and an accelerator grid (down-stream) is widely used. The screen grid is near cathode potential and directs the ions into small beamlets. A large negative potential is then applied to the accelerator grid to accelerate these ion beamlets out of the source. The sample is usually at ground potential, although this need not be the case. For example, negative sample biasing can be used to produce positive ion beam bombardment at energies which are larger ($E \sim 10$ keV) than most broad-beam ion sources are designed to handle.

The ability to extract high current (~ 1 mA/cm^2) at moderate voltages (~ 500–1000 eV) requires that the screen and accelerator grids be closely spaced with many small apertures in each. The 15 cm Ion Tech source, for example, uses two 1 mm thick pyrolytic graphite plates, each drilled with 576 small-diameter holes. The downstream holes (diameter ~ 0.5 mil) are somewhat smaller than the upstream holes (diameter ~ 0.8 mil) to maximize extracted ion current and minimize backstreaming of neutral species into the source. The plates are spaced about 1 mm apart and positioned so the respective holes align. Maintaining this small, uniform grid spacing over large areas (~ 100 cm^2) requires grid materials which are mechanically stable over a wide range of thermal conditions. Grids are most often made of materials such as Mo or pyrolytic graphite, although successful use of single-crystal silicon patterned by photolithography and anisotropic wet chemical etching has recently been reported (Kaufman et al., 1982b; Speidell et al., 1982). Such silicon grids have a high degree of precision, retain the uniform mechanical properties of single-crystal silicon, and are free of the usual mechanical disturbances (e.g., drilling) which accompany grid fabrication. Silicon grids may not be the best choice for certain reactive gas environments (e.g., O_2), although one could consider coating the grids with an electrically conductive, nonreactive layer.

The high positive ion density in the extracted beam can result in beam divergence due to Coulomb repulsion as well as surface charging of insulating samples, which can reduce anisotropy and uniformity of the etch. To minimize these effects, low-energy electrons from a hot filament are injected into the beam as it exits the source. These neutralizing electrons are readily distributed within the conducting plasma of the beam to give a uniform beam potential. A common misconception is that these electrons combine with the ions to produce a neutral beam. In fact, electron–ion recombination is not significant, and the beam consists of low-energy elec-

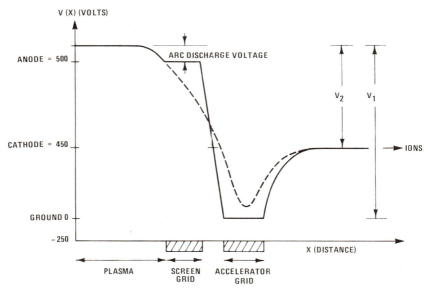

Fig. 12. Representative variation of electric potential through grids (solid line) and through apertures (dashed line) of a broad-beam ion source. Values shown are representative of 500 eV Ar^+ ion bombardment. (After Kaufman, 1978.)

trons and high-energy ions in a state of overall space-charge neutrality. Precise neutralization of the ion current is not necessary and a rough equality of electrons and ions is usually sufficient to maintain an insulating sample within a few volts of ground potential during ion bombardment.

The variation of electric potential through a broad-beam ion source is presented in fig. 12. Note that the ions originate at the plasma potential (the most positive potential) and are accelerated through a total voltage difference, V_1, between the plasma and accelerator grid. Since the plasma and screen grid potentials are nearly equal, V_1 is approximately the voltage between the two extraction grids. After extraction (de-celeration), the ion returns to ground potential with a net kinetic energy of eV_2. This so-called 'acceleration–deceleration' sequence (i.e., acceleration by potential V_1 and deceleration to potential V_2) serves two purposes. First of all, since V_2/V_1 is less than unity, a potential barrier is set up preventing neutralizing electrons from backstreaming into the source. Secondly, the extracted ion current density, j, is related to V_1 through a Child's Law-type expression of the form $j \sim V_1^{3/2}/L^2$, where L is taken to be the screen grid-to-accelerator grid spacing. Use of a large accelerator voltage, V_1, will therefore increase the ion current density at a fixed V_2, i.e., a fixed ion energy $= eV_2$.

3.3.2. Use of reactive gases in a broad-beam ion source

As noted earlier, Kaufman-type sources can be used with inert source gases (usually Ar^+) for inert ion beam etching or sputter deposition, or with reactive gases for RIBE or reactive sputter deposition. In fact, heavy ions such as $^{201}Hg^+$ were at first of greater interest than $^{40}Ar^+$ since space propulsion by ion momentum transfer was being pursued (Kaufman 1978). Specific issues which arise from the use of reactive gases in a broad-beam ion source to perform RIBE will now be addressed.

In general, the operation of hot filaments in reactive gas ambients, as compared to inert gas ambients, leads to reduced filament lifetime. In the Kaufman-type source described above, both the cathode and neutralizer were hot-filament thermionic electron sources (i.e., wires of tungsten, tantalum, etc.). We have found, as has Heath (1981a, 1982), that cathode lifetimes can be unacceptably short (about 10 h) when CCl_4 or CF_4 is fed into such an ion source. The cathode, being typically $\sim$40–80 V negative with respect to the plasma, erodes from sputtering due to ions in the discharge plasma. This would be true even in an inert ambient. With reactive gases such as oxygen in the source, filament erosion can be enhanced by chemical etching. Other effects such as the embrittlement of Ta wire in O_2 or N_2 discharges can lead to catastrophic filament failure from thermal or mechanical shock. Similar considerations apply to the neutralizer wire(s) as well, whose lifetime is also reduced by sputtering due to the extracted ion beam and by operation in reactive gases, although in this case the reactive gas pressure is probably an order of magnitude less than in the source. Although the conducting nature of the ion-beam plasma allows a neutralizer wire to be successfully used in a variety of positions, it must be immersed to some extent in the ion beam. Hence sputtering and chemical erosion due to reactive species diffusing through the extraction grids cannot be totally avoided. Also, one has the possibility of sputtered material from the wire contaminating the wafer surface downstream. The need to change filaments can lead to downtimes which are unacceptable in semiconductor processing fab lines. Also, whenever a RIBE or other plasma-etching system is exposed to room air, water vapor is adsorbed on the chamber walls. This is particularly troublesome when the system has been used to etch Al or Al alloys, since Al salts which remain on the chamber walls are hygroscopic. After pumpdown, this can lead to a high partial pressure of oxygen in the system and reduce RIBE etch rates due to Al_2O_3 formation (Downey et al. 1981).

Fortunately, cathodes are now available which are more suitable than hot filaments for RIBE application. For example, a hollow cathode has been incorporated in the so-called 'plasma bridge neutralizer (PBN)' which presents an alternate to the immersed-wire method of neutralizing the ion beam (Reader et al. 1978). Low-energy electrons are obtained by passing a flow of inert gas, e.g., Ar, through a small cylindrical chamber containing a biased emitter. An inert plasma discharge is then established in the chamber and the plasma extracted through a small hole. By biasing the PBN negative (~ 20 V) with respect to the beam, electrical coupling is facilitated. The name PBN arises since coupling between ion beam and neutralizer occurs via a connecting plasma plume or 'bridge'. Since the neutralizer can be physically located outside the ion beam, ion bombardment of the neutralizer does not occur, nor does contamination from neutralizer material sputtered downstream.

A hollow cathode (HC) can also be used to replace the hot-filament cathode in the discharge chamber. In this case, electrons extracted from a small inert gas discharge are used to establish a reactive gas discharge in the body of the larger RIBE ion source. One version, shown in fig. 13, uses a sintered tungsten insert impregnated with a low-work-function material as a thermionic cathode. An inert gas such as Ar is introduced

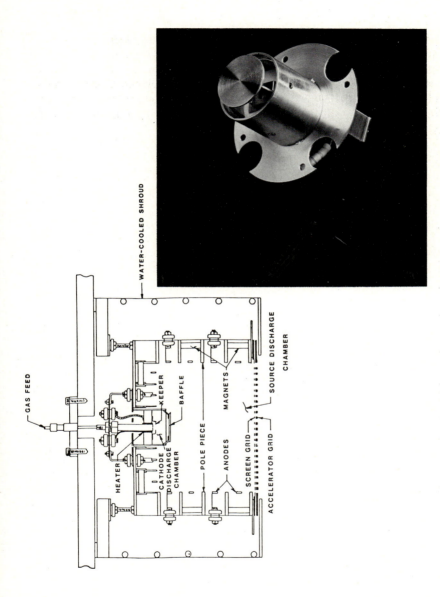

Fig. 13. Hollow cathode electron source (Ion Tech., Ft. Collins, CO).

down the barrel of the HC, where a discharge is initiated by heating the cathode tip to ~1000°C and applying about 350–500 V to the keeper (anode) electrode. A baffle partially isolates the HC discharge chamber from the source. As a result, a potential difference approximately equal to the argon discharge voltage exists across the annular aperture formed by the baffle and cathode body. Electrons are accelerated across this region into the source. Having energies in the range where ionization cross-sections for most gases are large, these electrons then establish a discharge in the source where a suitable reactive gas (CCl_4, CF_4, O_2, etc.) has been introduced. Since the HC never sees any reactive gases, hundreds of hours of reliable operation can be obtained.

By replacing both neutralizer and ion source filaments with long-lived, hollow cathodes, broad-beam ion sources can be constructed which are much more suitable for RIBE applications.

In addition to reducing the lifetime of hot filaments, reactive gases can also form polymeric materials on the ion-source walls, which degrade source performance and etch repeatability. For example, the ion current from a conventional Kaufman-type ion source run on fluorinated hydrocarbons such as C_2F_6, C_3F_8 or CHF_3 will decrease in time as insulating polymers deposit on the source walls. One worker has reported that after only 30 min of operation, a current density $\gtrsim 0.4$ mA/cm^2 could no longer be reliably maintained (Heath, 1981a, 1982). Other gases, such as CCl_4, can form conductive polymers or carbon flakes which may short out insulators or lodge themselves between the extraction grids. Frequent cleaning of the ion source, by abrasion or chemical means, is time-consuming and necessitates cycling the RIBE system between atmosphere and vacuum with the attendant problems of water vapor adsorption, operator exposure to noxious gases, etc.

One way of alleviating many of these problems is to use a conventional Kaufman-type ion source in which the anode, grids, and cathode cover plates are made of graphite. Such all-graphite sources (e.g., the Commonwealth Scientific Millatron G-Gun) reportedly can be run stably and continuously with fluorocarbon gases on the order of 100 h without the need for cleaning (Thompson, 1981). From a practical standpoint, this is of little advantage if filament lifetimes are only a few hours, and a long-lived, hollow cathode should be considered for use in such a source as well. Our own experience with CCl_4 in a conventional Kaufman-type source indicates that carbon plates out onto the source walls creating, in effect, an all-graphite source.

As noted in the introduction, one way to perform RIBE without directly introducing reactive gases into the ion source is to run the source on an inert gas and introduce the reactive gas into the process chamber instead (inert ion-assisted chemical etching). The Ar^+ ion-beam-assisted reactive gas-surface chemistry can produce favorable etch results in some cases. In addition, reactive gas molecules can diffuse to the acceleration grid where ionization occurs by collisions between inert gas ions and reactive gas molecules. The reactive gas ions so produced are then accelerated into the processing chamber, along with the inert ion beam.

If instead of backfilling the etch chamber with reactive gas, reactive gas is introduced at the sample surface by a directed gas jet in close proximity, a high effective pressure of reactive gas species is achieved at the sample surface, while most of the chamber remains at low pressure. Since the ion source discharge is essentially an inert gas discharge, source stability and filament lifetimes are not degraded by the reactive gases being used for etching in the chamber. By a proper choice of reactive gas species and ion flux/energy, high etch rates and good selectivity have been obtained for some materials (Geis et al., 1981; Chinn et al., 1983a, b). By varying the relative spatial orientation of neutral and ionic beams, a variety of topologically complex etch profiles can be tailored.

3.3.3. *The effect of a magnetic field*

As mentioned earlier, a magnetic field is used to alter the trajectory of primary electrons in the source so as to increase the probability of ionizing collisions with the source gas. Depending on source design, the field can be produced by electromagnets or permanent magnets. Permanent magnets tend to be less expensive than electromagnets, and since the field is fixed, do not require monitoring during processing. Electromagnets allow the operator to vary the magnetic field strength. Since electromagnets are sensitive to heat and cooling, process monitoring is needed to maintain the desired characteristics of the magnetic field. In this case, the operator may then be able to vary the **B** field strength and, by so doing, vary the beam intensity as well. When monatomic inert gases such as Ar, Xe, etc., are used in the source, beam *composition* is expected to be relatively insensitive to changes in **B**. However, most reactive gases of interest when using RIBE are molecular in nature: CF_4, CCl_4, CCl_2F_2, Cl_2, CHF_3, etc. The fragmentation pattern of molecular gases can be affected by the energy and density of electrons in the plasma discharge, which in turn is influenced by the applied **B** field (Mayer and Barker, 1981, 1982a). By this reasoning, one expects changes in B to affect the composition of both ionic and neutral species which exit the source.

Measurements of both the ionic and neutral components of the beam exiting a Kaufman-type source as a function of axial B-field strength have been reported by Mayer and Barker (1982a) for CF_4. Their results show that fragmentation of the parent gas into CF_x^+ ions and CF_y neutrals is

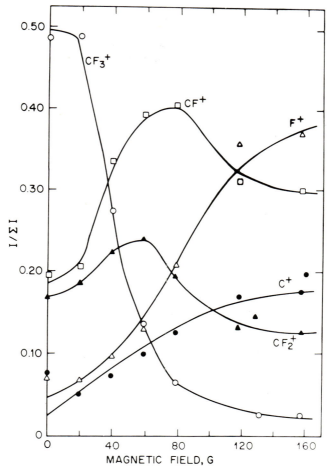

Fig. 14. Composition of ion beam as a function of axial magnetic field strength extracted from a CF_4 discharge (Mayer and Barker, 1982a).

extensive and, as expected, dependent on magnetic field strength. Figure 14 shows the variation of ion beam composition they measured as a function of axial B field strength up to $B = 160\,\text{G}$. Figure 15 shows the results for neutral species. In both cases, chamber pressure is $\sim 8 \times 10^{-5}\,\text{Torr}$ and source pressure is estimated at $\sim 8 \times 10^{-4}\,\text{Torr}$. A quadrupole mass analyzer was used to obtain the data.

As fig. 14 shows, fragmentation increases strongly with field strength. When B is low, CF_3^+ predominates; however, by $160\,\text{G}$, the populations of the lighter mass species C^+, F^+, CF^+ and CF_2^+ have increased greatly, while that of CF_3^+ has decreased by over an order of magnitude. Presumably this reflects the fact that at higher B field, energetic electrons have

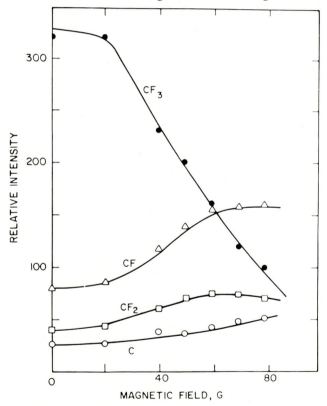

Fig. 15. Composition of neutral species extracted from CF_4 discharge as a function of axial magnetic field strength (Mayer and Barker, 1982a).

longer residence times in the source, and therefore multiple electron impacts on ions are more likely. One implication of this for RIBE is that the etch rate and selectivity will depend on magnetic field as well, since the individual ionic species comprising the beam will in general behave differently. For example, etching of Si at 500 eV with a mass-selected ion beam derived from CF_4 has shown that F^+ beams etch Si four times faster than CF^+ (Mayer and Barker, 1982b). It remains to be seen whether a variable ***B*** field can profitably be used in RIBE to tailor selectivity or increase etch rates without compromising other etch parameters such as uniformity. In addition, this approach may be less effective in a multipole-configured ion source since most of the source volume where ion production occurs is ***B*** field free.

3.3.4. Use of low energy ions

Using dual-grid extraction optics, broad-beam electron bombardment ion sources can easily produce current densities of 0.5 mA/cm^2 at 500 eV. At much lower energies, however, the extracted current density is space-charge limited to impractically low values. For example, at 100 eV, only about one tenth the value at 500 eV can be extracted, i.e., $\sim 0.05 \text{ mA/cm}^2$. Data of Harper et al. (1981) show, however, that ion energy does not have to be high to activate reactive gas-surface chemistry, so that RIBE can be effectively carried out at low energy ($\sim$100–200 eV). There are several reasons why one might want to perform RIBE at such low ion energies. Working at ion energies where physical sputtering is minimal can reduce the possibility of device damage penetration of contaminants during RIBE etching. Other sputtering artefacts such as trenching and redeposition, which can result in sloped sidewall profiles, may also be less pronounced. Harper et al. (1981) measured the etch rates of Si and SiO_2 with ion beams derived from Ar and CF_4 over a wide energy range of 20–1500 eV and investigated the effectiveness of ion bombardment in enhancing chemical etching. They found that the etch rates of CF_4/Si and Ar/Si were very similar, suggesting that physical sputtering, by Ar^+ or CF_3^+, dominates the etch process in this case. On the other hand, the etching yield (atoms/ion) of SiO_2 using CF_4 was significantly higher than with Ar. This difference in etch yield, attributed to ion-enhanced chemical etching, increased with ion energy up to $\sim$200 eV and then remained constant. Hence by RIBE etching SiO_2 in CF_4 at energies $\lesssim 200$ eV, the enhancement of chemical etching by ion bombardment was optimized, while unwanted effects caused by physical sputtering were reduced.

The usuable ion energy range for broad-beam ion sources has recently been extended to low energies by using a single extraction grid of very fine aperture size in place of the double-grid ion optics (Harper et al., 1981). In order to overcome the space-charge limitation of a double grid, the aperture size in the single grid is chosen to be comparable to or less than the plasma sheath thickness (≤ 0.1 mm). Sources designed with single grids have demonstrated high ion current density (~ 1 mA/cm^2) at energies below 100 eV without sacrificing either tight beam collimation or small energy spread (~ 10 eV). A major disadvantage of the single-grid approach is that there is no screen grid to deflect ions away from the accelerator grid. This results in increased sputter erosion of the grid and reduced lifetime. Using a 100 mesh grid (about 40 holes/cm) of electroformed Ni mesh, an Ar$^+$ ion current density of 1 mA/cm^2 was reported for ion energies in the 20–100 eV range. Grid lifetime was several hours (Harper et al., 1981). Use of energies ≥ 200 eV is expected to reduce grid lifetime significantly due to increased sputter erosion and grid overheating.

3.3.5. Unconventional ion sources

Although Kaufman ion sources are widely used to perform RIBE today, they are by no means the only sources. Several alternatives have been explored for RIBE applications, including: a modified magnetron ion gun (Okano and Horiike, 1982) and a cold-cathode ion source of the saddle-field type (Rovell and Goldspink, 1981). A broad-beam ion source using a microwave electron cyclotron resonance (ECR) discharge has been gaining in popularity with Japanese workers (Miyamura et al., 1982; Matsuo and Adachi, 1982; Miyamura et al., 1983; Watakabe et al., 1983; Matsuo, 1983; Asakawa and Sugata, 1983) and, as of this writing, an ECR-based RIBE system is commercially available from Anelva Corporation (Tsukada, 1983). Only a brief discussion of these sources will be presented here, and the reader is referred to the work of the Japanese workers mentioned above for more details.

Broad-beam ion sources for RIBE are generally operated with a DC plasma discharge, as compared to the RF frequencies used in plasma and RIE reactors which are typically in the range of 40 kHz to 20 MHz. Nevertheless, workers have reported plasma etching at much higher frequencies, and recently a broad-beam ion source has been developed for RIBE in which ions are extracted from a microwave electron cyclotron resonance (ECR) discharge at 2.45 GHz (Matsuo and Adachi, 1982; Miyamura et al., 1982). Figure 16 shows the basic design. Microwave

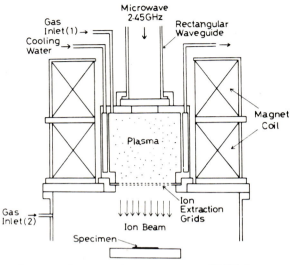

Fig. 16. 15 cm diameter electron cyclotron resonance (ECR) ion source (Matsuo and Adachi, 1982).

power is coupled to the discharge chamber through a rectangular waveguide with a fused quartz window. The discharge chamber functions as a microwave cavity resonator. The high magnetic field (about 875 G) is produced by water-cooled coils surrounding the source and produces the condition for cyclotron resonance for the electrons in the source. Ion extraction utilizes a dual-grid design similar to that in a Kaufman-type source. One notable difference is that the accelerator grid is at ground potential while the whole discharge chamber is floated at the extraction voltage (0–2 eV). Since no cathode, thermionic or otherwise, is used in the source, degrading of cathode stability or lifetime in the reactive gas ambient is not a problem. Current densities of 1 mA/cm² can be drawn from the ECR source with uniformity better than ±10% over a 13 cm diameter and ±5% over a 1 cm diameter. Results of RIBE etching SiO_2, Si, and Al with ions extracted from C_4F_8 and $SiCl_4$ plasmas in an ECR source are shown in fig. 17. Highly selective etching of SiO_2 : Si in C_4F_8 was obtained with etch rate ratios ~20 : 1. Miyamura et al. (1982) have reported selectivities as high as 30 : 1 with C_4F_8 in an ECR source (1 keV ion energy).

Although magnetron discharges have been used for sputter deposition for many years, the use of cold-cathode-type magnetron ion guns (MIG)

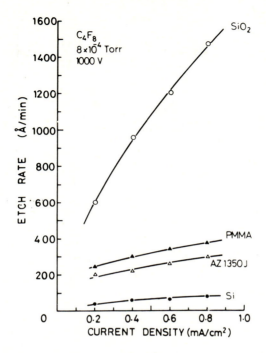

Fig. 17. Etch rate of various materials for 1 keV ions extracted from a C_4F_8 discharge in an ECR ion source (Matsuo and Adachi, 1982).

to perform RIBE is relatively new (Okano and Horiike, 1982). As originally designed, a large and unwieldy electromagnetic coil assembly was required to compensate for magnetic flux leakage around the cathode, which reduced the discharge intensity. In a more recent and improved design (shown in fig. 18), the magnetic coil has been replaced by a permanent rare-earth Sm–Co magnet located at the center of the cathode. This more compact configuration produces a large magnetic field (~ 0.4 Wb/cm^2) across the narrow annular gap (~ 2 mm) surrounding the cathode, since flux leakage is minimal. In this design, the iron cathode acts as a magnetic pole piece. A DC plasma discharge is established by electron impact ionization in the gap region through $\boldsymbol{E} \times \boldsymbol{B}$ electron field ($\boldsymbol{E}$ is the electric field between anode and cathode, and $\boldsymbol{B}$ is the field in the gap). Reactive ions of energy equal to the discharge voltage are extracted directly from the plasma. The MIG has been used to RIBE etch SiO_2 and

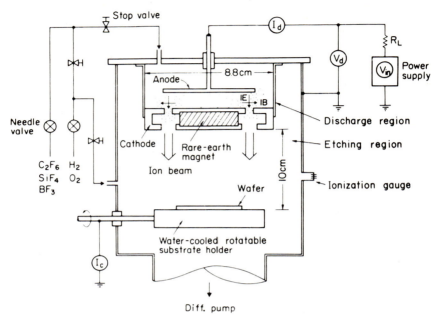

Fig. 18. Schematic diagram of RIBE system based on magnetron ion gun source (Okano and Horiike, 1982).

undoped poly-Si with C_2F_6, SiF_4, and BF_3 gases introduced directly into the source. Maximum etch rates for these gas/substrate combinations at 500 eV were around 1000–2000 Å/min. Using C_2F_6, 1 μm patterns could reportedly be etched in SiO_2 without undercutting of the PMMA resist (Okano and Horiike, 1982). Furthermore, the use of C_2F_6 gave selective etching of SiO_2 with respect to Si (etch rate ratio of SiO_2/poly-Si ≃ 10/1). Addition of H_2 *outside* the ion gun gave an etch ratio of SiO_2/poly-Si as high as 25. Using Kaufman-type sources, RIBE etching of SiO_2 and poly-Si gives selectivities of about 6–7 : 1, which are comparable to the MIG results when C_2F_6 was used without H_2. Using chlorosilane ($SiCl_4$), Al could be etched approximately six times faster than SiO_2, although ion energies below about 500 eV reduced the Al etch rate dramatically. In all cases, etch profiles were sufficiently anisotropic to permit faithful submicron pattern transfer.

Lastly, a cold-cathode source of the saddle-field configuration has been used for RIBE with CF_4 and CHF_3 source gases (Rovell and Goldspink,

1981). The plasma chamber, anodes, and cathode grid of the ion source (Ion-Tech, Ltd., Middlesex, England) are all made of graphite. During operation, the source is water cooled, but the anodes reach temperatures as high as 900°C. Ion current densities from this source are not particularly high (~ 0.1 mA/cm^2) and etch uniformity may be a concern for IC applications. (The authors suggested rotating and moving the target radially using a planetary system.) Furthermore, at energies typically used in RIBE, $\lesssim$ 500 eV, the saddle-field source pressure is very high (10^{-1} Torr) and etch rates for Si and SiO$_2$ are only about 100–200 Å/min. For all of the above reasons, this RIBE source does not appear to be as promising an alternative to Kaufman-type sources as the magnetron or ECR-based sources.

4. Mechanisms of RIBE etching

Progress in our fundamental understanding of RIBE has principally come from controlled ion beam studies of:

(1) reactive ion sputtering (e.g., CF$_3^+$ → Si, SiO$_2$);

(2) inert ion sputtering in the presence of reactive gases (e.g., Ar$^+$ + Cl$_2$ → Si).

Moreover, such studies, involving directed ion beams with independently variable energy, flux, incidence angle, etc., allow fundamental aspects of plasma-assisted etching to be explored to a degree not usually possible in systems such as RIE and plasma reactors in which the plasma parameters are not as easy to vary independently or to quantify. A basic understanding of RIBE etch mechanisms, therefore, will be of value in understanding the behavior of more complex dry etch systems. In a similar vein, an understanding of physical sputtering has been of value in formulating a theory of chemically enhanced sputtering applicable to RIBE (Mayer and Barker, 1982b).

As noted earlier, the name 'Reactive Ion Beam Etching' is a simplification and does not accurately describe the technique as it is usually practiced. In addition to the beam of energetic reactive ions which bombard the sample, both neutral reactive species (e.g., diffusing downstream from the ion sources) and ambient gas molecules are also present at the sample surface and may participate in the etch process. Since chamber pressures are normally $\sim 1 \times 10^{-4}$ Torr, collective impingement rates of these molecules are $\simeq 10^{16}$ cm^{-2} s^{-1}, which are comparable to the flux of reactive ions (1 mA/cm^2 of singly-charged ions corresponds to 6×10^{15} ions/cm^2 s). In general, therefore, one has to deal with a variety of interactions which,

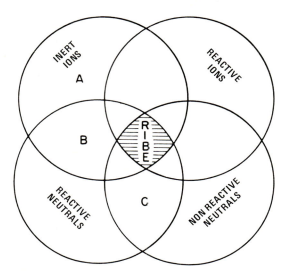

Fig. 19. Venn diagram illustrating the interplay of ionic and neutral species in RIBE.

taken collectively, determine the etch rate, selectivity, anisotropy, etc., of the RIBE process.

This is schematically indicated by the Venn diagram in fig. 19, where the darkened area represents RIBE in its most general form. Directional ion bombardment of the substrate involves inert species (e.g., Ar^+) and/or reactive species (e.g., Cl_2^+, CF_3^+). Neutral species are identified as either nonreactive or reactive with the substrate (in the absence of the ion beam). For a silicon substrate, nonreactive neutrals might be F_2, CCl_4, N_2, Ar, etc. Reactive neutrals might be free radicals such as $Cl^.$ or molecular ambient gases such as O_2 or Cl_2. Ion bombardment will in general affect the degree of 'reactivity' of both types of neutral species with the substrate. Directional ion milling (physical sputtering alone) is represented by region A for an inert ion beam. In the case where inert ion milling is carried out in the presence of a reactive gas, region B is applicable. Examples include the use of added oxygen to increase the selectivity of a physically sputtered Ti or Al metal mask, or the use of a high Cl_2 partial pressure at a GaAs surface to increase its etch rate under Ar^+ bombardment (Geis et al., 1981). If ion bombardment is absent and dry etching occurs only by chemical formation of volatile species, region C is applicable. In a RIBE configuration, this could be simulated by placing the sample close to the source extraction grids without, however, extracting an ion beam. The sample would then see only the background gas and reactive and nonreactive neutral species diffusing from the plasma across the pressure gradient (~ 0.2–1×10^{-3} Torr/cm) from ion source to etch chamber.

In order to arrive at a qualitative understanding of the mechanisms underlying RIBE, it is useful to consider separately the effects of: (1) ion-beam-enhanced gas–surface chemistry, (2) reactive ion bombardment, and (3) reactive neutrals on the etch process.

4.1. Ion-beam-enhanced gas–surface chemistry

Ion–surface interactions in the context of plasma-assisted dry etching have been studied for many years. It has long been known, for example, that an incident ion beam can enhance (suppress) the reactivity of neutral gas species adsorbed at a sample surface leading to etch rates significantly greater (lower) than would be obtained in the presence of the neutral species alone. A well-known example by Coburn and Winters (1979) is the case of Ar^+ sputtering Si in the presence of XeF_2. Under either Ar^+ bombardment (450 eV) or XeF_2 exposure alone, Si etch rates were low ($\lesssim$ 5 Å/min). When ion bombardment and XeF_2 exposure were performed simultaneously, however, Si removal rates were over ten times greater (~ 55 Å/min). Another particularly striking example is the etching of GaAs, which under Ar^+ bombardment (500 eV, 1 mA/cm^2) is ~ 2500 Å/min. When a jet of molecular Cl_2 is simultaneously directed at the sample, rates as high as 40 000 Å/min have been reported (Geis et al., 1981)! Other examples include ion beam etching of Si, III–V compounds or refractory metals in the presence of Cl_2 gas (Geis et al., 1981; Chinn et al., 1983a). In cases where ion-assisted gas–surface chemistry leads to nonvolatile product formation, suppressed etch rates can result – as in the case of $Ar^+ \rightarrow$ Al in the presence of F_2 (Coburn and Winters, 1979). Finally, there are cases where the addition of an ion beam has little or no effect on the etch rate, such as the case of $Ar^+ \rightarrow$ Al in the presence of Cl for which no ion-enhanced chemistry is observed (Winters, 1983). The reader is referred to Mayer and Barker (1982b), Coburn and Winters (1979), Gerlach-Meyer (1981), Tu et al. (1981), Barker et al. (1983), Haring et al. (1982), Winters (1983), Winters et al. (1983), Tachi et al. (1982) and Mayer et al. (1981) for more detailed discussions of ion-assisted gas–surface chemistry.

One way to understand the role of the incident ion beam (or any directed radiation source, for that matter) is to consider how it interferes with or enhances one or more of the steps in the total reaction chain resulting in dry etching; viz, chemisorption, formation of a product molecule, and desorption of the product molecule. For example, if Ar^+ bombardment increases the number of possible chemisorption sites on the sample surface, or aids in the removal of reaction products (stimulated desorption), an enhanced etch rate may occur. If on the other hand the ion beam dissociates adsorbed fluorocarbon gases and deposits carbon on the surface, etching may be suppressed due to scavenging of reactive species by the carbon and/or due to its low physical sputter yield. And so on. In some cases, the concept of a reaction chain may not be applicable and other models need to be developed. For example, Haring et al. (1982) have suggested that the etching reaction of Si under Ar^+ bombardment in the presence of XeF_2 proceeds stepwise and that the enhanced etch rate observed is not caused by enhanced production of a final, volatile end product molecule (SiF_4). Instead, weakly-bound intermediate products (SiF_x) may be formed at the surface, which are readily sputtered away by the Ar^+ beam.

Even in the absence of a detailed understanding of the ion-assisted gas-surface chemistry occurring during RIBE, general etch trends can sometimes be predicted. A good example is the etching of Si by Ar^+ in the presence of Cl_2. Several papers have recently explored the interaction between inert-gas ions and reactive neutrals such as Cl_2 or XeF_2 on a Si surface (Mayer and Barker, 1982b; Gerlach-Meyer, 1981; Tu et al., 1981; Barker et al., 1983; Haring et al., 1982). One model, developed by Gerlach-Meyer (1981), explains ion-induced etching by local excitation of adsorbed species at the surface. This excited state can then react to form volatile product molecules which desorb. Excellent agreement was obtained between the calculated and measured dependences of the flow rate of either XeF_2 or Cl_2 on the etch yield of Si under inert ion (Ar^+, Ne^+, Ae^+) bombardment. Figure 20 shows results for the case of $Ar^+ + XeF_2 \rightarrow Si$ (Gerlach-Meyer, 1981). The silicon etch rate is taken to be a separable sum of three terms: the spontaneous etch rate in the presence of XeF_2 (not affected by ion current), the physical sputtering rate (assumed not to be affected by XeF_2 gas flux), and the ion-induced etch rate, ER_i. The latter term was calculated by using an expression of the approximate form:

$$ER_i \simeq \frac{BJZ}{(2A + 2BJ + Z)}, \tag{1}$$

where Z = incident XeF_2 gas flux, J = incident inert-ion flux (ions/cm²s),

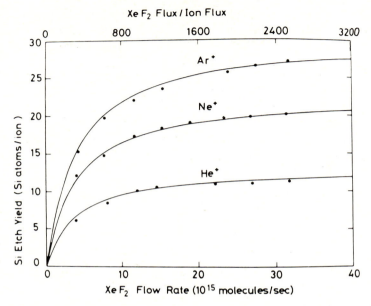

Fig. 20. Etch rate of Si versus ratio of XeF$_2$ gas flux to ion flux of Ar$^+$, Ne$^+$ and He$^+$. Dots are experimental data; solid lines are calculated. (Gerlach-Meyer, 1981.)

B = the number of surface molecules excited by one bombarding ion, and A^{-1} represents the 'strength' of the ion-induced etch process.

Both A and B are empirically fitted parameters with values $A = 4.7 \times 10^{15} \, \text{s}^{-2} \, \text{cm}^{-2}$ and B = 30 (Ar$^+$), 22 (Ne$^+$), 13 (He$^+$), appropriate to the curves drawn in fig. 20. Although the effectiveness of the ion-induced etch mechanism was not influenced by using different ions (i.e., same value of A obtained), the number of excited surface molecules (value of B) did depend on the incident ion species, increasing with ion mass.

Another treatment of the Ar$^+$ + Cl$_2$ → Si system, proposed by Barker et al. (1983), is based on simple considerations of mass balance for the species participating in the etch process. This approach (see Mayer and Barker, 1982b, Barker et al., 1983 for a detailed treatment), is sufficiently general to be of use in other cases and we briefly outline it below.

Figure 21 represents the Si surface under molecular Cl$_2$ collisions and Ar$^+$ ion bombardment (ion flux = J). The number of collisions/cm^2 s of neutral Cl$_2$ is given by $Z = P/(2\pi mkT)^{1/2}$, where m and P are the Cl$_2$ mass

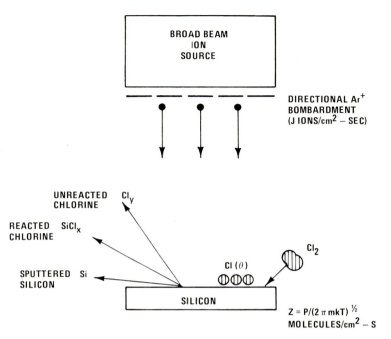

Fig. 21. Representation of Si surface under simultaneous Ar$^+$ ion bombardment and molecular Cl$_2$ collisions.

and partial pressure at temperature T. The surface coverage of Cl atoms = θ where $0 < \theta < 1$. We assume that Cl$_2$ adsorption goes as $(1 - \theta)$, with a sticking probability S. Taking into account the available adsorption sites, the flux of Cl *atoms* (two per molecule) being deposited on the Si surface is $2SZ(1 - \theta)$.

Adsorbed chlorine is assumed lost from the surface by ion-stimulated desorption. Each ion collision is assumed an isolated event leading to a subsequent desorption, so that total desorption rates can be obtained by summing over single ion events. The Ar$^+$ beam desorbs Cl in both unreacted (e.g., Cl$_y$) and reacted (e.g., SiCl$_x$) forms. If the effective cross sections for these processes are, respectively, α and β, then the ion-induced loss of Cl is given by

$$\alpha y \theta J + \beta x \theta J = (\alpha y + \beta x) \theta J.$$

In steady state, the mass balance equation for Cl gives

$$2SZ(1 - \theta) = (\alpha y + \beta x)\,\theta J.$$

The surface coverage of Cl is then calculated to be:

$$\theta = (1 + \phi J/Z)^{-1}, \tag{2}$$

where $\phi = (\alpha y + \beta x)/2S$.

Silicon is removed from the surface in both elemental (unreacted) and chloride (reacted) form. Assuming for convenience that the reacted product molecule is $SiCl_4$, the etch rate R of Si is given by:

$$R = (1/4)\beta\theta J + Y(1 - \theta)J, \tag{3}$$

where Y is the physical sputtering yield of exposed Si at the surface by Ar^+.

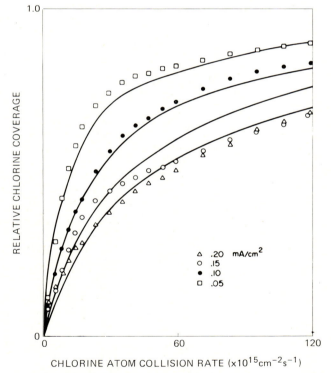

Fig. 22. Experimental chlorine surface coverage (θ) versus chlorine atom flux (J) for several Ar^+ ion beam current densities (Barker et al., 1983). The solid lines plot $\theta = (1 + \phi J/Z)^{-1}$ from eq. (2) in the text.

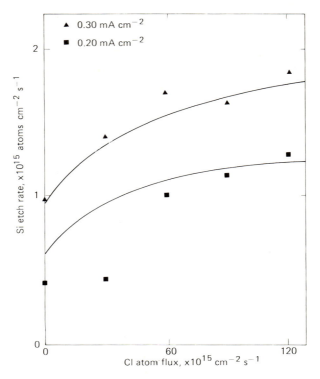

Fig. 23. Ar$^+$ ion-assisted etch rate of Si vs. neutral flux of Cl for two Ar$^+$ ion current densities. Solid curves are calculated for 600 eV Ar$^+$ energy (Barker et al., 1983).

Equation (3) clearly shows (within this model at least) the way in which physical sputtering (right-hand term) and ion-beam-assisted gas–surface chemistry via SiCl$_4$ desorption (left-hand term) both contribute to the measured dry etch rate. In addition, gas adsorption affects the magnitude of both terms. As coverage (θ) increases, more product chloride molecules can be formed and desorbed, but consequently, less elemental Si is exposed to be sputtered away.

Since the etch rate of Si as a function of Cl$_2$ pressure and Ar$^+$ ion current density can be measured experimentally and the coverage of chlorine determined, e.g., by using Auger spectroscopy, it is straightforward to test the qualitative predictions of eqs. (2) and (3) and obtain best-fit values for the parameters α, β. For the Ar$^+$ + Cl$_2$ system, this simple approach does in fact successfully predict the surface coverage of reactive gas molecules in the presence of ion and neutral collisions as well as the observed dependence of etch rate on both Ar^+ ion and neutral Cl_2 fluxes (see figs. 22 and 23) (Barker et al., 1983).

Behavior similar to that observed for $Ar^+ + Cl_2 \rightarrow Si$ has been reported by Geis et al. (1981) for inert ion beam etching of GaAs in the presence of molecular Cl_2, and by J.D. Chinn et al. (1983a) for GaAs and Ti. In addition, these latter workers explored the effect of sputtering with a reactive Cl_2^+ beam in the presence of Cl_2. The enhanced etch rates observed were directly related to the concentration of chlorine at the sample surface, whether incident from the Cl_2^+ beam or from the background gas. Etching proceeded via the formation of volatile chlorides such as $GaCl_3$, $AsCl_3$ and $TiCl_4$.

4.2. Reactive ion bombardment

In the previous section, ion-induced gas–surface interactions were discussed. If the ion beam is inert, the ion–surface interaction in the absence of adsorbed gases is described by physical sputtering. That is, the collision cascade produced by momentum transfer from incident ion to substrate atoms results in some atoms near the surface gaining sufficient energy (several eV) and momentum to escape into vacuum. Sputtering yields for a variety of materials under Ar^+ bombardment (500 eV, 1 mA/cm^2) typically range from about 0.5–2.5 atoms/incident ions. If the ion beam is reactive, however, the chemical reaction between ion and target can significantly alter the target sputter yield. Such chemically enhanced physical sputtering (reactive sputtering) is the counterpart of physically enhanced gas–surface chemistry described earlier (e.g., $Ar^+ + Cl_2 \rightarrow Si$).

The reactive ion beam, then, not only causes particle ejection by momentum transfer, but delivers potentially reactive chemical species to the surface. The formation of volatile or low-binding-energy species can enhance etch rates over that of purely physical sputtering where several eV are required to escape from the surface; the formation of nonvolatile or tightly bound species can suppress the etch rate. In either case, the use of a reactive ion beam adds a whole new dimension to sputter etching.

Figure 24 shows the yield of SiO_2 in ion beams derived from Ar or CF_4 discharges (Mayer et al., 1981). The CF_4/SiO_2 reactive sputter yield is seen to be greater than the inert Ar/SiO_2 physical sputter yield at all beam energies employed, due to the formation of volatile reaction products in the

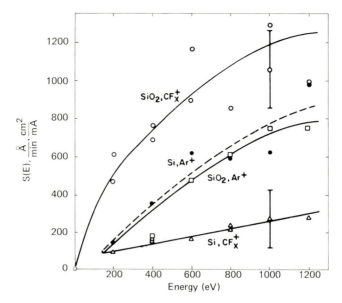

Fig. 24. Etch yield of SiO_2 and Si in ion beams derived from Ar or CF_4 discharges. Beam angle of incidence = 30°. (After Mayer et al., 1981.)

case of CF_4. The enhancement is greatest at lower energies $\lesssim 200$ eV, where the reactive yield (1.8 atoms/ion) is many times greater than that of Ar^+ (~6 times greater at 200 eV).

Since the reactive ions extracted from a CF_4 plasma are in general of the form CF_x^+, the reactive sputter rate shown in fig. 24 actually represents the collective etch rates for each separate fluorocarbon ion. The reaction probabilities of CF_x^+ species with Si and SiO_2 have been investigated by several groups (Mayer and Barker, 1982a, b; Miyake et al., 1982). The relative reactivities of these with Si and SiO_2 are tabulated in table 2 (see Mayer and Barker, 1982b). While CF_3^+ is seen to be the most effective ion for etching SiO_2 at this energy, the same is not true for Si. Differences such as these probably reflect differences in the surface binding energy of new chemical species created by the reactive beam. For example, stable C–O and Si–F bonds formed at the SiO_2 surface in the presence of CF_x^+ bombardment can lead to loosely bound species such as CO, CO_2, SiF_4 and enhanced etch rates, while CF_x^+ bombardment of Si has been shown to produce a protective carbonaceous polymer layer which impedes the etch pro-

Table 2
Relative reactivities of 500 eV CF_x^+ ions with SiO_2 and Si.[a]

Ion	SiO_2 (Mayer and Barker, 1982a)	Si (Miyake et al., 1982)
CF_3^+	6.1	0.45
CF_2^+	3.2	0.32
CF^+	1.8	0.15
C^+	0.4	–
F^+	0.4	0.60

[a]Mayer and Barker, 1982b.

cess (Coburn and Winters, 1979). Table 2 shows that the highest reactivity of CF_x^+ with Si occurs when no carbon is brought to the surface at all (i.e., ion beam is pure F^+).

In addition to differences in etch rate between physical sputtering and reactive sputtering, the dependence of etch rate on ion incidence angle can also be quite different. Figure 25 shows the etch yield $S(\theta)$ of Si under Cl^+ bombardment for incidence angles from $\theta = 0$ (normal) to $\theta = 60°$. The yield peaks at normal incidence for all energies shown (400–1200 eV) and is only weakly dependent on angle (Mayer et al., 1981).

The physical sputtering yield is qualitatively quite different from the reactive yield in that it rises slowly with increasing incidence angle (rather than decreasing) and peaks for $\theta \sim 60°$ (rather than at $\theta = 0$). This behavior is similar for many other materials under inert ion bombardment (maximum yield for $\theta \sim 40–60°$) with some exceptions – such as gold whose sputter yield behaves much more like the reactive sputter yields in fig. 25. This angular dependence of yield is fairly representative of other reactive gas–substrate combinations for which $S(\theta)$ has been measured – e.g., $CF_4 \rightarrow SiO_2$, $CF_4 \rightarrow$ photoresist (Matsui et al., 1980; Okano and Horiike, 1981).

Mayer and Barker (1982b) have chosen to view the $S(\theta)$ vs θ curve for reactive ion bombardment as being enhanced at small angles (near normal incidence) relative to the curve for inert ion bombardment, but unchanged at large angles (grazing incidence). They conjecture that the reason for this modification relates to both surface damage and reactive ion implantation caused by the ion beam, which in turn foster the formation of new chemical bonds with surface atoms. If the surface atom binding energies are low enough, the sputter yield is enhanced. These effects should be largest at normal incidence. At grazing angles, much less lattice disorder is created

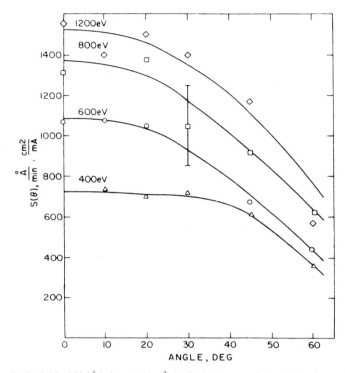

Fig. 25. Etch yield of Si (Å/min · mA/cm²) in Cl_2 ion beams (400–1200 eV) as a function of incidence angle (after Mayer et al., 1981).

and fewer ions are implanted. Chemical bond formation is affected less and the beam is expected to behave as if it were inert with regard to the angular dependence of sputter yield.

In the previous discussion, we have considered etching with inert ion beams where the ambient gas is potentially reactive in the presence of the ion beam (e.g., $Ar^+ + F_2 \rightarrow Si$), as well as etching with reactive ion beams (e.g., $CF_3^+ \rightarrow Si$). A further variation is to bombard the substrate with a reactive ion beam in the presence of a reactive gas. This situation is a close approximation to the way in which RIBE is usually performed. In this regard, Mayer and Barker (1982a) have analyzed the case of $CF_x^+ + CF_y$ (neutral flux) $\rightarrow SiO_2$ using the simple mass balance model described earlier for $Ar^+ + Cl_2 \rightarrow Si$. Again, the model does not depend explicitly on the

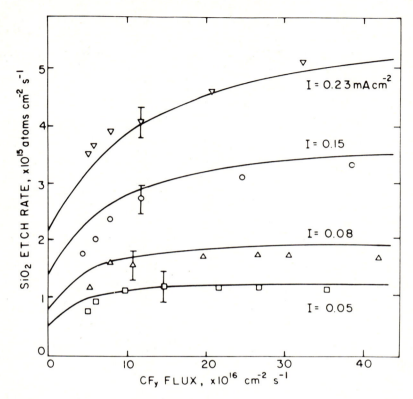

Fig. 26. Etch rate of SiO_2 by CF_x^+ as a function of CF_y neutral flux (Mayer and Barker, 1982b). The solid lines are calculated after eq. (4) in the text.

detailed mechanisms for energy transfer and reaction enhancement at the surface. Instead, the major assumption is that the energy required to promote chemical reactions leading to volatile product formation/desorption (i.e., dry etching) is supplied by the kinetic energy in the ion beam. Figure 26 shows that good agreement could be obtained with the experimental rate by using an expression of the form

$$\text{Etch Rate} \equiv R \equiv \frac{\beta J(SZ + J)}{SZ + (\alpha + \beta)J}, \qquad (4)$$

involving ion current density (J), neutral gas pressure P ($Z = P/[2\pi mkT]^{1/2}$) and three fitting parameters S, α and β, whose values in fig. 26 were chosen to be $S = 0.15$, $\alpha = 1.6$ and $\beta = 1.3$. At low currents and high pressure, the etch rate becomes proportional to ion current density ($R \to \beta J$), in agreement with the experimental curves. As the adsorbate coverage saturates, additional neutral flux has little effect and the rate is dominated by ion-beam-induced desorption. At very low pressure, $R \to \beta J/(\alpha + \beta)$. For an inert beam, this rate would simply be that for physical sputtering.

4.3. Neutral species

Neutral species contribute to the RIBE process in a variety of ways. These species may be gases purposely introduced as a nondirectional background into the etching chamber, directed via gas jet at the sample surface, or introduced from behind the sample (e.g., Ar or He for gas-assisted wafer cooling). They also include the usual residual gases (CO, H_2, H_2O) found in any diffusion-pumped or cryopumped UHV system. Finally, they can result from gases introduced directly into the ion source, which either diffuse chemically unaltered through the source extraction grids or are changed/fragmented into other neutral species within the plasma discharge before drifting into the etch chamber.

One way in which neutral species can affect the RIBE etch is by participating in charge exchange collisions with the incident ion beam when the ions and atoms are like species. Resonant charge exchange, which has a particularly large cross-section, can result in energetic neutral bombardment of the surface – a neutral flux which will *not* be read on a current probe at the sample. Consider, for example, ion-beam-assisted etching with a 500 eV Ar^+ beam. The mean free path for Ar^+–Ar resonant charge exchange collisions is ~130 cm at 10^{-4} Torr (Harper et al., 1982). In many cases of practical interest in RIBE, however, the directed ion is some fragment of the original molecule, so that resonant charge exchange is only possible with a small percentage of the background gas. For example, ions extracted from a CF_4 plasma are principally CF_3^+, CF_2^+ ..., while the background is mainly CF_4. Without CF_4^+ ions present, ion-neutral charge exchange interactions during beam transport to the sample are not very likely.

As discussed earlier, the synergistic effect of ion beam and neutral gas fluxes at the sample surface is central to the RIBE process. Unfortunately, the chemical makeup of the neutral species emanating from the ion source is, in general, complex. In addition to the source gas, a variety of fragmented molecular species can appear in the etch chamber background.

Barker and coworkers have measured the flux of neutral species produced by CF_4 in a Kaufman-type source as a function of both axial magnetic field in the source and chamber pressure (Mayer and Barker, 1982b). At low fields ($\lesssim 10\,G$) or high pressures ($\gtrsim 2 \times 10^{-4}\,Torr$), CF_3 species dominate. At higher fields (~ 40–$80\,G$) or lower pressures ($\lesssim 1 \times 10^{-4}\,Torr$), the neutral population contains significant percentages of CF_3, CF_2, and C. F and F_2 were expected, but for experimental reasons were difficult to detect.

When chemically reactive neutral species are present at the sample surface, RIBE etch rate, etch profiles, and overall etch quality can all be affected. For example, RIBE of GaAs in CCl_4 has shown that the etch rate decreases with sample distance from the ion source for a given incident ion flux (Powell, 1982). When the source-to-sample distance was increased about 30%, from 15 to 22 cm, the RIBE etch rate decreased by about 30%. Presumably this behavior reflects the decreasing density profile of reactive free radicals (Cl?) as one goes downstream from the ion source. A similar variation of etch rate with distance has also been observed in RIBE of Al-CuSi alloys using CCl_4, where the etch rate dependence was found to decrease linearly with source-to-substrate distance (Downey et al., 1981), although in this case the decrease was more gradual (e.g., a 200% increase in distance decreased the etch rate by only about 60%).

The flux of diffusing reactive neutral species at the sample surface is much less directional than the ion beam. As more 'chemistry' is added to the RIBE process via reactive neutral species, the etch can become increasingly isotropic. This gives one the ability to tailor sidewall profiles by adding more or less chemically reactive species to the etch process. For example, Chinn et al. (1983b) have demonstrated that RIBE-etched wall profiles of Si can be either overcut, vertical, or undercut, depending on the background pressure of highly reactive XeF_2 gas. XeF_2 behaves as a source of atomic fluorine (F^-–Xe^{++}–F^+ decomposes readily), which spontaneously etches Si. Under 500 eV Ar^+ ion bombardment, an overcut slope characteristic of physical ion milling is obtained. As the partial pressure of XeF_2 in the etch chamber is increased, the profile becomes more vertical, eventually obtaining an undercut slope characteristic of isotropic chemical etching.

As a last example of the effect of reactive neutrals on RIBE etching, we note the etching of AlCuSi alloys in CCl_4 (treated in detail in section 5.1). In this case, the source-to-substrate distance was varied (i.e., population of reactive neutrals was varied) until the chemical removal rate of Al via volatile $AlCl_x$ formation matched the sputter removal rate of non-

Table 3
Summary of etch scenarios discussed in section 4.

Etch description	Example(s)
(1) Ground state chemical reactions	$F \to Si$
	$Cl \to Al$
(2) Chemically enhanced physical sputtering	
(a) reactive ions	$Cl^+ \to Si$
(b) reactive ions with background gases which	$Cl^+ + F \to Si$
are/are not reactive without beam	$Cl^+ + Cl \to Si$
(3) RIBE	$Cl^+ + F + Cl + Ar^+ \to Si$
(4) Physically enhanced chemical reactions	
(a) no reaction without beam	$Ar^+ + Cl \to Si$
(b) reaction without beam	$Ar^+ + F \to Al$
(5) Physical sputtering (inert ion beam milling)	$Ar^+ \to Si$

volatile $CuCl_x$. Even in the absence of any reactive neutrals, the etch would proceed via chemically-enhanced sputtering from the Cl_2^+, Cl^+ ion beam. However, in this case, the ability to fine-tune the etch process by adjusting the neutral population allowed one to optimize the overall quality of the etch (Downey et al., 1981).

Table 3 is a summary of the etch scenarios discussed in this section.

Cases (5) and (1) represent, respectively, material removal by physical sputtering and spontaneous chemical formation of volatile species in the absence of external activators (electron or ion bombardment, UV radiation, higher than ambient temperature, etc.). Intermediate cases (2)–(4) represent etching with a greater 'physical component' than case (1) and a greater 'chemical component' than case (5). Ion-beam-assisted etching is represented by (2) for the case of reactive ion beam bombardment and by (4) for the case of inert ion bombardment. In either case, ion bombardment can be performed in the presence of neutral species which either do or do not react strongly with the substrate without ion bombardment. RIBE in its most general form (3) involves ion bombardment (reactive and inert) in the presence of neutral species whose reactivity with the substrate is altered in the presence of the directed energetic radiation (ions and photons) from the ion source.

5. Patterning and etching materials with RIBE

Although as a commercially available technique, RIBE has only been around for approximately five years, most materials of interest to Si and III–V technology have reportedly been etched. This includes silicon, polysilicon, SiO_2, Si_3N_4, Al and its alloys, refractory metals and metal silicides, photoresist and polyimides, GaAs $Al_xGa_{1-x}As$, and InP. Nevertheless, compared to reported applications of dry etching in general, applications of RIBE are still in an early stage of exploration. (Of all the published papers dealing with dry etching, those on RIBE probably account for less than 5%.) In this section, we review results on RIBE etching of a variety of materials.

5.1. Aluminum and aluminum alloys

Aluminum and its alloys are now, and are expected to remain, an important metallization for both interconnects and contacts to Si. To avoid spiking problems, Si (1%) is commonly alloyed to aluminum (Merchant, 1982, D'Heurle and Ho, 1978). Shrinking device geometries have resulted in increased interconnect lengths, reduced linewidths and the need for metallizations to carry larger current densities. This, in turn, has increased the possibility of electromigration-induced failures in AlSi lines. One way of reducing electromigration in Al is by reducing grain-boundary diffusion by the addition of elements such as Cu, Ni, Cr and Mg. The most commonly used of these is copper, at concentrations typically in the 2–4% range. Unfortunately, the addition of this much copper to aluminum creates additional etching problems when chlorine-containing plasmas are used to etch aluminum. Namely, at normal etching temperatures, cuprous chloride is nonvolatile. In order to remove the cuprous chloride, the wafer has to be heated to temperatures > 200°C, where lateral attack of the aluminum has been observed (Schaible and Schwartz, 1979). Another way to remove the cuprous chloride is by adding a sputtering component to the etch process. This technique works well and eliminates the lateral etch problem. Other problems encountered in the etching of aluminum and its alloys are removal of aluminum oxide from the surface of the wafers to be etched, and post-etch corrosion resulting from water vapor reacting with surface chlorine or cuprous chloride, which subsequently attacks the metallization. To date, there has been limited success with etching Al alloys such as AlCu(4%)Si using traditional dry etching techniques. Plasma and RIE

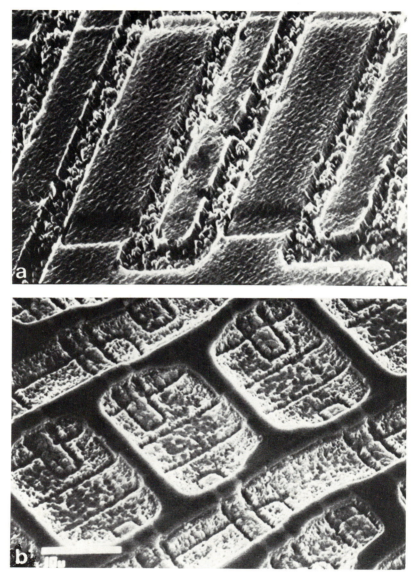

Fig. 27. RIBE etch of AlCuSi with incorrect balance of chemical and physical components; (a) chemistry-rich, (b) physics-rich with sputtering artifacts. (Reprinted with permission from Solid State Technology, Technical Publishing, a company of Dun & Bradstreet.)

etching of high copper-content AlCuSi alloys generally result in an etch residue which is hygroscopic. When the residue is exposed to humidity, corrosion in the interconnect metallization can result. Argon ion milling of these alloys results in redeposition and trenching, which also seriously compromises yield.

Etching of AlCuSi in CCl_4 using RIBE has been accomplished by an appropriate combination of reactive chemistry and physical mechanisms (Downey et al., 1981). Use of too much chemistry has been found to replicate the disadvantages of reactive plasma etching, while too much of the physical component can produce sputtering artifacts and reduced selectivity (figs. 27a and 27b, respectively).

The effect of low-energy active neutrals on the etching of AlCuSi has been reported earlier (Downey et al., 1981). It was found that at very short source-to-substrate distances (<17.8 cm), the etch quality is unacceptable, with incomplete removal of the Si and Cu compounds. As source-to-substrate distance is reduced, the system approaches a chemistry-rich mode in which the etch process is dominated primarily by active, low-energy neutrals. The etch rate data indicated that the density of active neutrals decreased linearly with distance.

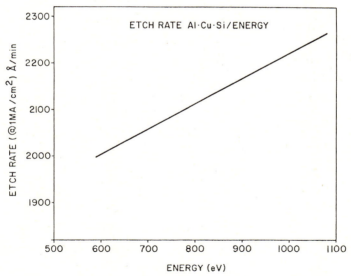

Fig. 28. Dependence of AlCuSi etch rate on energy of ions extracted from a CCl_4 discharge (Downey et al., 1981).

The dependence of the AlCuSi etch quality on ion beam energy has also been reported (refer to fig. 28). The etch rate is seen to increase slightly with increasing ion energy. At energies below approximately 625 eV, copper and silicon residues were left. Although above 625 eV, the copper residues were fully removed, energies greater than 675 eV were required before the Si was fully removed. At 775 eV, some sputtering artifacts begin to appear. By operating at ion energies between 675 eV and 775 eV, however, high copper-content (4%) AlCuSi alloys could successfully be

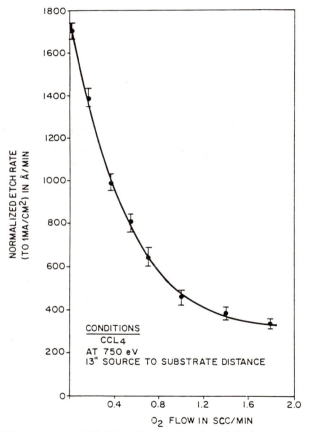

Fig. 29. RIBE etch rate of AlCuSi in CCl_4 as a function of O_2 flow into the etch chamber (Downey et al., 1981; reprinted with permission from Solid State Technology, Technical Publishing, a company of Dun & Bradstreet).

etched. Source-to-substrate distance required for the optimum etch quality was 21.6 cm.

The effects of oxygen on the etching of aluminum wafers in a plasma etch system have been previously reported (Chapman and Nowack, 1980, Tokunaga and Hess, 1980). Similar effects of oxygen on the etching of AlCuSi have been observed in RIBE using CCl_4 at 750 eV with a 33 cm source-to-substrate distance (Downey et al., 1981). Oxygen was introduced into the etch chamber by a controlled leak through a mass flow controller. All etch rate measurements were obtained by determining the time required to etch a 1000 Å layer of AlCuSi. The results are illustrated in fig. 29. The normalized etch rate with no oxygen flow (i.e., with only the residual oxygen of the vacuum system present) is 1700 Å/min. As the oxygen flow is increased, the etch rate falls off dramatically to 340 Å/min at a flow of 1.79 sccm. This reduction in etch rate is attributed to the increasing rate of formation of Al_2O_3 on the surface of the wafer as the partial pressure of oxygen in the system increases. At very high oxygen

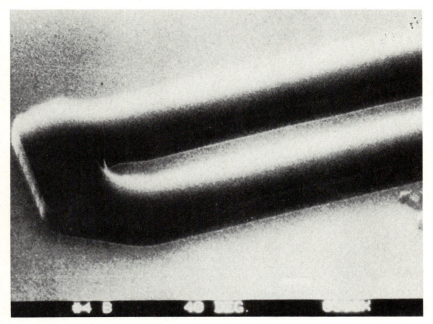

Fig. 30. RIBE etch of AlCu(4%)Si(1.5%) metallization pattern in CCl_4 with photoresist remaining. SEM magnification = 8500×. (Reprinted with permission from Solid State Technology, Technical Publishing, a company of Dun & Bradstreet.)

levels (e.g., immediately after cycling from atmosphere to vacuum), no measurable etch rate could be observed. During venting, water vapor is absorbed by the hygroscopic etch residue and acts as a source of residual oxygen during subsequent etching. This suggests the use of a vacuum-load-lock wafer-handling system and cryopumping for reproducible, production-worthy etching of aluminum and its alloys.

The best RIBE etch results reported for AlCuSi have been obtained using CCl_4 at 700–715 V, with a source-to-substrate distance of 21.6 cm and maintaining good control over the oxygen and H_2O content of the vacuum chamber and load-locks (refer to figs. 30–32). Under these conditions, etch rates up to 2200 Å/min with 3–4 : 1 selectivity for standard photoresists (e.g., AZ1350-J and HPR-204) have been reproducibly obtained. At these ion energies, the selectivity with respect to SiO_2 is consistently greater than 2 to 1. Although this selectivity is not high, the reactive ion beam is quite uniform (approximately ±5% over a 100 mm wafer) and

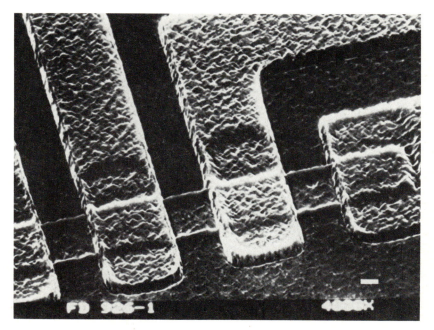

Fig. 31. RIBE etch of AlCu(4%)Si(1.5%) pattern after photoresist mask was stripped. SEM magnification = 4800×.

Fig. 32. AlCu(4%)Si(1.5%) (a) before RIBE with CCl_4 and (b) after RIBE and stripping of AZ1350J resist.

an instantaneous electrical end-point detector (Downey et al., 1981) was able to provide precise and uniform stopping at the underlying substrate.

As illustrated in fig. 33, etching of AlCuSi is a multi-step process:

(1) The chemically inert, native aluminum oxide layer is sputtered away and the metal surface is continually cleared of aluminum oxide grown during the remaining etch by the sputtering component of the beam. The amount of Al_2O_3 to be etched is dependent upon the oxygen in the system; the rate of its removal depends on the energy and current density of the ions employed.

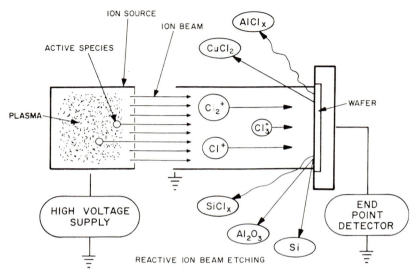

Fig. 33. Schematic representation of RIBE etching AlCuSi alloy with Cl_2^+, Cl^+ ions.

(2) Aluminum is etched chemically, and the volatile chlorides are subsequently pumped away.

(3) Copper chemically reacts to form nonvolatile chlorides which are sputtered away. In plasma and RIE etch systems, these copper chlorides typically remain on the wafer. When exposed to humidity, the copper chlorides react to form cuprous oxide and hydrochloric acid, causing rapid post-etch corrosion. The ion energy required in the standard configuration described above to remove the nonvolatile copper chlorides is >625 eV.

(4) Silicon is removed by a combination of physical and radiation-enhanced chemical processes. In the standard configuration, ions of energy greater than 675 eV are required to remove all the Si residue at the AlCuSi/SiO_2 interface. Below 675 eV, Si speckles are left behind, similar to those observed in the wet chemical etching of AlCuSi alloys.

When all of these mechanisms occur in the correct proportions, a clean anisotropic etch with no trenching, no redeposition, and minimal post-etch corrosion has been obtained with RIBE.

5.2. Silicon dioxide and silicon

VLSI technology has created the need to etch aluminum and its alloys with high resolution and anisotropy. Similar requirements must be met for silicon dioxide and silicon as well. This section addresses the results of reactive ion beam etching of these materials in either Cl or F-containing etch gases. As a preview, the results to be discussed in this section are summarized in table 4.

5.2.1. RIBE with chlorine-containing gases

Both silicon dioxide and polysilicon have been successfully RIBE etched using CCl_4 (Downey et al., 1981). At an ion energy of 600 eV and a beam

Table 4
Comparison of reported results of RIBE etching SiO_2 and Si.

Etchant		Energy (eV)	J (mA/cm^2)	Rate (Å/min)	Reference
CCl_4	SiO_2	600	0.60	1120	Downey et al.,
	Si	750	0.80	1398	1981
$Ar + CCl_4$	SiO_2	1000	5.3×10^{-3}, 10^{15} mol/s cm^2	~0.8 at/ion	Gerlach-Meyer
	Si	1000	5.3×10^{-3}, 10^{15} mol/s cm^2	~0.5 at/ion	et al., 1980
CF_4	SiO_2	500	0.1	~2.5 at/ion	Harper et al.,
	Si	500	0.1	0.5 at/ion	1981
CF_4	SiO_2	800	0.4	~500	Heath, 1982
	Si	800	0.4	~100	
CF_4	SiO_2	800	(Normalized) 1.0	~1000	Mayer and
	Si	800	(Normalized) 1.0	~200	Barker, 1981
C_2F_6	SiO_2	900	0.40	~650	Heath, 1982
	Si	900	0.40	~100	
C_2F_6	SiO_2	900	0.40	~650	Brown et al.,
	Si	900	0.40	~65	1980
C_2F_6	SiO_2	600	Not given	~2000	Okano and
	Si	600	Not given	~200	Horiike, 1982
C_3F_8	SiO_2	900	0.40	~650	Heath, 1982
	Si	900	0.40	~100	
Cl_2	SiO_2	800	(Normalized) 1.0	400	Mayer and
	Si	800	(Normalized) 1.0	1000	Barker, 1981

current density of $0.60\,mA/cm^2$, silicon dioxide was etched with an etch rate of $1120\,\text{Å}/min$. Because SiO_2 is an insulator, a hot wire filament neutralizer was used to avoid charge buildup. Figure 34a shows an etch of a $2800\,\text{Å}$ layer of thermal oxide on bulk silicon before the photoresist was removed. The mask is $1.1\,\mu m$ of tapered HPR-204. Figure 34b is the same etch after the removal of the photoresist. Figure 34c is a cross-sectional view of the same etch with the photoresist stripped. Note that the photoresist which was tapered prior to etch has transferred its taper into the etched oxide.

Polysilicon has also been etched in CCl_4 at an energy of $750\,eV$ and a beam current density of $0.80\,mA/cm^2$ with an etch rate of $1400\,\text{Å}/min$ (Downey et al., 1981). Again, an electrical end-point detection scheme was employed to provide excellent stopping on the underlying SiO_2. Profilometer and ellipsometric measurements indicate that the etch was stop-

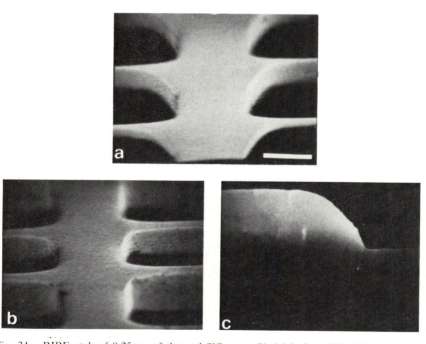

Fig. 34. RIBE etch of $0.25\,\mu m$ of thermal SiO_2 over Si: (a) before HPR-204 photoresist stripping; (b, c) after resist strip. The resist was intentionally tapered before dry etching. (Reprinted with permission from Solid State Technology, Technical Publishing, a company of Dun & Bradstreet.)

ped well within 200 Å of the underlying SiO_2 with excellent uniformity over the entire wafer. In fig. 35 is shown an etch of 3750 Å of n-doped polysilicon over 2800 Å of thermal SiO_2 patterned with 1.1 μm of HPR-204 positive photoresist. It is evident from fig. 35 that there is no undercutting, no trenching and no redeposition, and that the sloped photoresist has left the polysilicon with an even taper.

Gerlach-Meyer et al. (1980) have investigated the etching of SiO_2 and Si using inert ion beams of argon with a background gas of CCl_4. In fig. 36a, the etch rate for Si is shown as a function of time. For times < 3.2 min, the etch rate indicated is for argon ion beams alone; after 3.2 min, the CCl_4 background gas is turned on. The immediate decrease in etch rate is a result of the chemisorption of the CCl_4 molecules on the surface. Once steady-state is attained, the overall etch rate has decreased by 16%. Figure 36b shows the corresponding curve for SiO_2 versus time. In this case, after the CCl_4 is turned on, there is a 70% increase in the SiO_2 etch rate when steady-state is attained. This work demonstrates that adding a

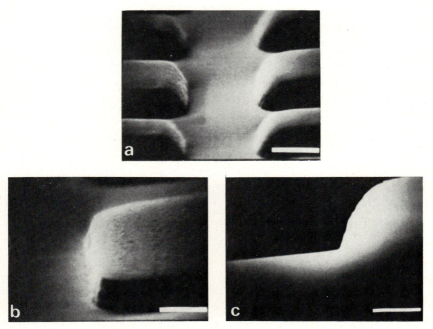

Fig. 35. RIBE etch of 0.375 μm of n-doped polysilicon over thermal SiO_2: (a) before strip of HPR-204 resist; (b, c) after resist strip.

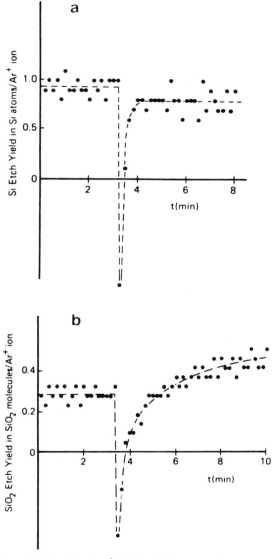

Fig. 36. Ar$^+$ etching (1 keV, 5.3 μA/cm^2) yield of (a) Si and (b) SiO$_2$ with and without a CCl$_4$ flux, Q. $Q = 0$ for $t < 3.2$ min; $Q \sim 10^{15}$ molecules/cm^2/s for $t > 3.2$ min (Gerlach-Meyer et al., 1980).

properly chosen reactive gas to what was purely physical sputtering can increase selectivity of the process – in this case, from 1 : 1 to 2 : 1.

Si and SiO$_2$ have also been etched in ion beams of Cl$^+$ and Cl$_2^+$ by introducing Cl$_2$ directly into an ion source (Mayer et al., 1981). In fig. 37 is shown the etch rate so obtained for Si and SiO$_2$ versus ion energy with samples at a 30° angle of incidence to the beam. The etching of Si at low energies was very slow and required an etch at 1000 eV for 30 s to remove the native oxide layer which inhibited the etching. The data depicted in fig. 37 for Si is compensated for the 30 s etch required to remove the native oxide layer. We see that the etch rate of Si in Cl$_2$ is enhanced over that obtained for physical sputtering. The dependence of etch rate versus angle for Si and SiO$_2$ is shown in fig. 24 (section 4.2). The curve for SiO$_2$ exhibits the profile expected for physical sputtering, while that for Si is representative of chemically dominated etching up to the highest energies studied. For a 0° angle of incidence (i.e., normal incidence), the etch rate ratio of Si to SiO$_2$ at 600 eV is about 5 : 1.

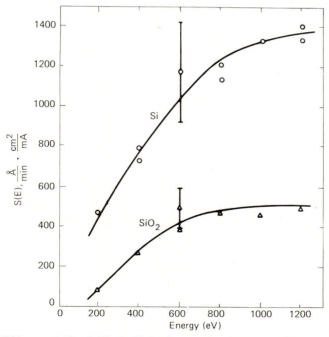

Fig. 37. Etch rate of Si and SiO$_2$ in Cl$_2$ ion beam versus ion energy (Mayer et al., 1981).

5.2.2. RIBE with fluorine-containing gases

Extensive research has been dedicated to the evaluation of the etching characteristics of SiO_2 and polysilicon with fluorine-containing gases. The main objective in many of these studies is the achievement of a high etch rate ratio of SiO_2 to Si for etching contact vias. Other applications, such as patterning a polysilicon or polycide gate electrode, require a low etch rate ratio of SiO_2 to Si, so the thin gate oxide is not consumed.

The etching of silicon dioxide and polysilicon has been investigated using ion beams generated from CF_4, C_3F_8 and C_2F_6 plasmas (Heath, 1982; Brown et al., 1980; Heath 1981b). Figure 38 shows the etch rates of polysilicon and SiO_2 versus ion energy for CF_4 and Ar at $0.40\,mA/cm^2$ ion beam current density. SiO_2 is seen to etch several times faster in CF_4 than in argon. Polysilicon, on the other hand, etches at a slower rate in CF_4 than with argon for energies below 800 eV. Above 800 eV, the etch rate of polysilicon with CF_4 rises linearly with energy to a value approximately

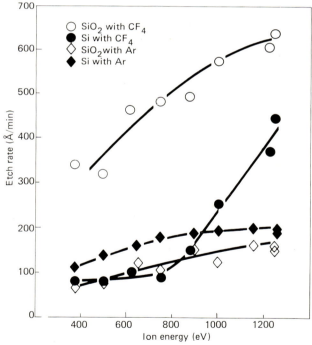

Fig. 38. Etch rate versus ion energy using Ar or CF_4 for SiO_2 and polysilicon (Heath, 1982).

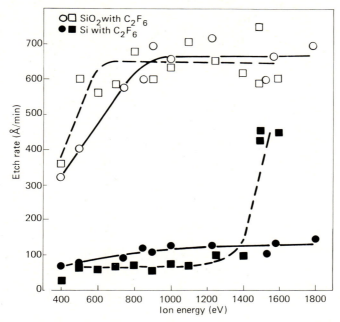

Fig. 39. Etch rate versus ion energy using C_2F_6 for SiO_2 and polysilicon. Squares are untrapped diffusion-pumped system. Circles are cryopumped system. (Heath, 1982.)

twice that with argon at an energy of 1200 eV. The etch rate ratio of SiO_2 to Si is 5:1 at lower energies, falling off rapidly at energies above 800 eV.

The etch rate versus energy of polysilicon and SiO_2 is shown in fig. 39 for the etchant C_2F_6 whose C-to-F ratio is 30% greater than in CF_4. The beam current density again was held constant at $0.40 \, mA/cm^2$. Two sets of etch rates are shown: one obtained in a cryopumped system and the other from a diffusion-pumped system. The etch rates for SiO_2 are comparable in the two systems, increasing linearly with energy and reaching a plateau at higher energies. The etch rate of the polysilicon is different in the two systems. In the diffusion-pumped system, the etch rate is lower and shows a rapid rise at 1400 eV. This results in differing etch selectivity between the two systems. At energies below 1200 eV, the selectivity in the diffusion-pumped system is 10:1, as compared to 6–7:1 for the cryopumped system. Presumably these differences result from more carbon contamination in the diffusion-pump system, where hydrocarbon oil can backstream into the etch chamber.

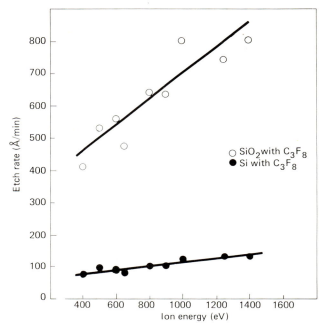

Fig. 40. Etch rate versus ion energy for SiO_2 (open circles) and polysilicon (closed circles) in C_3F_8 (Heath, 1982).

Figure 40 shows the etch rates of polysilicon and SiO_2 in C_3F_8. Polysilicon etches very similarly in C_3F_8 as in C_2F_6. The SiO_2 etch rate, however, does not level off, as seen for C_2F_6. The selectivity of SiO_2 to Si in C_3F_8 is 6:1 over the energy range investigated.

In fig. 41, the etch rates in C_2F_6 of various types of silicon dioxide (thermal oxide grown in steam, Nitrox SiO_2, and LPCVD SiO_2) and polysilicon (undoped and heavily doped) are compared. The etch rates of doped and undoped polysilicon are essentially the same, unlike the case of plasma etching. The etch rates of steam and Nitrox SiO_2 again are quite similar, while that for LPCVD SiO_2 is higher.

The etching of SiO_2 and Si using ion beams generated from a CF_4 plasma has also been studied by Harper et al. (1981). They extended the lower limit of the ion energy spectrum down to 20 V by employing an ion source with a single extraction grid. This work again shows an enhancement of the RIBE etch rate of SiO_2 in CF_4 over that for argon ion milling and a suppression of the etch rate for silicon (fig. 42).

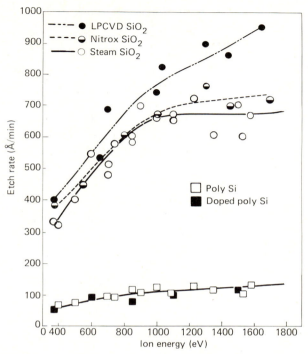

Fig. 41. Etch rate versus ion energy using C_2F_6 for doped and undoped polysilicon and SiO_2 deposited by various techniques (Heath, 1982).

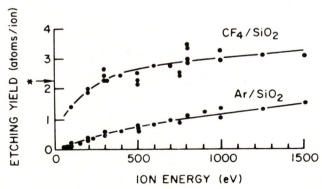

Fig. 42. Etching yield (atoms/ion) versus ion energy for CF_4 or Ar on Si and SiO_2 (Harper et al., 1981).

Mayer et al. (1981) have investigated the effects of angle of incidence, ion flux and background pressure on the etching of SiO_2 and polysilicon with ion beams generated from CF_4. Figure 24 (section 4.2, above) shows the etch rate of SiO_2 and polysilicon as a function of beam energy at constant background gas pressure and 30° angle of incidence. Again, enhanced etching of SiO_2 over that in argon is observed, while there is a suppression of the etch rate for polysilicon. Etch rates as a function of angle of incidence at various beam energies are shown in fig. 25. With argon, the physical sputtering rate peaks for incidence angles of approxi-

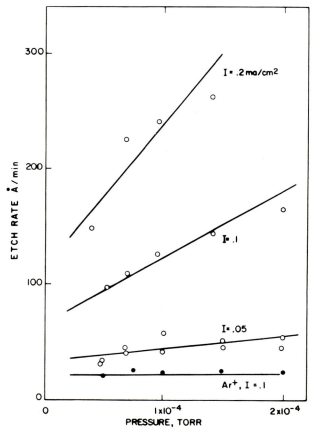

Fig. 43. Etch rate of SiO_2 as a function of background gas pressure using Ar^+ (0.1 mA/cm²) and CF_x^+ (0.05–0.2 mA/cm²) beams at 600 eV and 30° incidence angle (Mayer et al., 1981).

mately 45–60°, while in the case of CF_4 (at low energies), the peak occurs at normal incidence. At higher energies where sputtering by the reactive CF_x^+ beam is enhanced, the peak shifts to angles of about 30–40°, which are more typical of physical sputtering. Figure 43 is a plot of etch rate as a function of background pressure for constant-energy ion beams derived from CF_4 and argon. This data gives an indication of the importance of nonenergetic, potentially reactive neutrals to the etching process (discussed in detail in section 4). An interesting feature is the dependence of the etch rate slope on beam current.

Okano and Horiike (1982) have etched SiO_2 and polysilicon using C_2F_6, SiF_4 and BF_3. An unusual feature of this work was that the ion beams were generated from a magnetron ion gun. It was observed that the etch rate for SiO_2 increased linearly with increasing discharge current for all the gases used. For polysilicon, however, a linear increase in etch rate with discharge current was observed for SiF_4, BF_3 and argon, but not for C_2F_6. For C_2F_6, the etch rate is nearly constant with current and also has the lowest etch rate of all the gases studied. The selectivity of SiO_2 to Si is shown in fig. 44. When oxygen was added to C_2F_6, the etch rate of SiO_2 decreased rapidly and then became constant. The polysilicon etch rate, however, increases

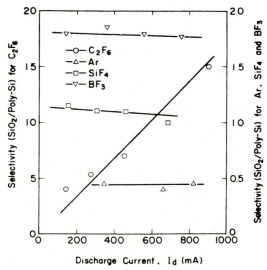

Fig. 44. SiO_2-to-polysilicon etch rate ratio as a function of discharge current. Left ordinate for C_2F_6; right ordinate for Ar, SiF_4 and BF_3. (Okano and Horiike, 1982.)

with increasing O_2 content, eventually exhibiting the same trends as SiO_2. With the addition of H_2 to C_2F_6, etch rates of both SiO_2 and Si decrease, with the polysilicon etch rate decreasing more rapidly. The overall selectivity value, however, increases dramatically to 25:1, which is among the highest reported for RIBE of SiO_2 over silicon.

5.2.3. Etching mechanisms for silicon and silicon dioxide

5.2.3.1 Chlorine-containing gases. The work of Gerlach-Meyer et al. (1980) (fig. 36, above) shows that the etch rate of polysilicon decreases with the addition of CCl_4 to argon, while the etch rate of SiO_2 increases with the addition of CCl_4. The work of Mayer et al. (1981) shows, however, that the etch rate of polysilicon increases in Cl_2 over that of argon and the etch rate is maximum at 0° (fig. 25, above). This indicates that Cl_2 chemically etches the polysilicon and that carbon retards the etch rate. The dependence of the etch rate for SiO_2 on angle of incidence indicates that the etching of SiO_2 proceeds by a combination of sputtering depending on the energy of the ion beam. The etch rate of SiO_2 in ion beams derived from Cl or Cl_2 is comparable to that of physical sputtering in argon, while using CCl_4, a 70% increase was noticed (Mayer et al., 1981). It appears that the presence of carbon on the surface enhances the etch rate of SiO_2 via the formation of volatile CO and CO_2.

We now present a simple picture of RIBE etching Si and SiO_2 in CCl_4 or Cl_2 which is consistent with the above results. This is meant only to give guidance in understanding the overall etch process and should not be taken as saying anything about detailed reaction mechanisms. It appears that the CCl_4 is chemisorbed on the Si surface. Dissociative chemisorption may then be enhanced by ion bombardment, producing a selective accumulation of carbon which, if not physically removed, retards the chemical reaction leading to $SiCl_4$ production. In the case of Cl and Cl_2, since there is no retarding layer of carbon present, the etch rate increases and the etch proceeds primarily through the formation of $SiCl_4$.

In etching SiO_2 with CCl_4 and Cl_2, it appears again that the dissociative chemisorption of CCl_4 is assisted by ion bombardment. The carbon is able to form volatile compounds with the oxygen in SiO_2, leaving the Si readily available to form the volatile $SiCl_4$. In the case of etching with chlorine, the carbon is unavailable to free the Si, and the etching must proceed primarily through physical mechanisms.

Based on the above observations, it appears that selective etching of Si over SiO_2 should occur best in a chlorinated gas containing no carbon, such as Cl_2 or BCl_3, etc. Addition of carbon is expected to retard the etch rate of Si and enhance the etch rate of SiO_2. A chlorinated gas, however, would not be the optimum choice for selective etching of SiO_2 over Si.

5.2.3.2. Fluorine-containing gases. It has been observed (Okano and Horiike, 1982; Mayer et al., 1981; Gerlach-Meyer et al., 1980; Brown et al., 1980; Heath, 1981b) that the etch rate of SiO_2 in fluorinated-hydrocarbon gases is enhanced over that of pure physical sputtering in argon, while the etch rate of Si is suppressed. The increased etch rate of SiO_2 is attributed to formation of volatile SiF_4 under ion bombardment. Harper et al. (1981) propose that this increase in the formation of volatile SiF_4 results from the formation of ion bombardment-induced active sites. Surface carbon is removed as the volatile species CO, CO_2, or COF_2 (Harper et al., 1981; Brown et al., 1980). At lower energies, Mayer et al. (1981) observed a maximum etch rate for SiO_2 at 0° angle of incidence, suggesting a chemical etch mechanism is operative. At higher energies, the maximum shifts up to 30–45°, indicating the addition of a physical component to the etch-

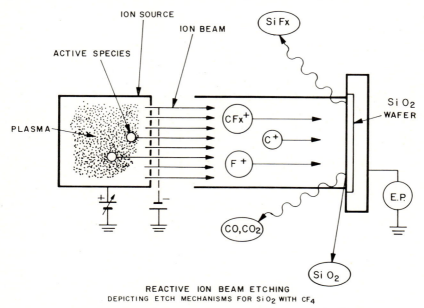

Fig. 45. Simplified representation of RIBE etching SiO_2 in fluorocarbon gases.

ing. The important effect which neutrals have in the enhancement of the etch rate is indicated by the data shown earlier in fig. 43. In fig. 45, a simple model for the etching of SiO_2 in fluorocarbon systems is sketched.

The suppression of the silicon etch rate in fluorocarbon gases has been attributed to the accumulation of carbon on the Si surface (Coburn and Winters, 1979; Mayer et al., 1981). This is also supported by the work of Heath (1982) where a decreased etch rate of Si was observed in a diffusion-pumped system (fig. 39) or when the C/F ratio of the etch gas was increased (figs. 38–40), and by the work of Okano and Horiike (1982) where increased etch rates of Si were obtained with noncarbon-containing gases BF_3 and SiF_4. In order to etch Si and form volatile SiF_4, the carbon must be removed from the surface. Unlike the case of SiO_2, no lattice oxygen is available to form volatile CO, CO_2 or COF_2, so that this carbon must be removed by sputtering or the addition of O_2 to the system. Figure 46 is a sketch of the proposed mechanisms of etching Si in a fluorocarbon system. The selective etching of SiO_2 to Si seems best addressed by use of a fluorocarbon with a high C/F ratio. Table 4 (section 5.2) summarizes the

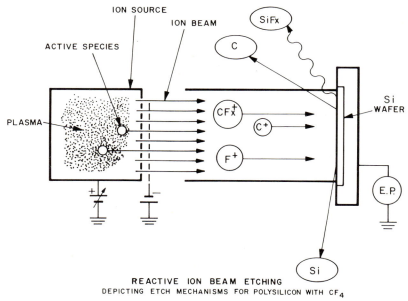

Fig. 46. Simplified representation of RIBE etching Si in fluorocarbon gases.

effect of various experimental parameters in RIBE etching SiO_2 and Si with general fluorocarbon gases, C_xF_y.

5.3. Photoresist and organics

It is interesting to note that the earliest commercial applications of both plasma etching and RIBE were concerned with the erosion of photoresist. In plasma etching, the goal was to find an alternative to wet chemical removal of photoresist (Tolliver, 1980). In place of organic photoresist strippers, chemicals such as acetone, or solutions such as chromic–sulphuric acid and sulphuric acid–hydrogen peroxide, it was found that an O_2 plasma in a barrel reactor could be used to quickly and safely strip photoresist from wafers or de-scum resist residue from areas to be subsequently etched. Photoresist molecules (C_xH_y) are converted into volatile species such as CO, CO_2 and H_2O by the active oxygen in the discharge. Such RF plasma 'photoresist ashers' were among the earliest commercial equipment to take advantage of dry etch technology (c. 1967).

In the case of RIBE, the situation was just the opposite of what developed in plasma etching. Namely, it was desired to minimize rather than enhance the erosion of organic masking materials which occurred under dry etching with inert ion beams. Inert ion beam etching does not, in general, exhibit high selectivity, with the result that the sputter rate of photoresist is comparable to that of Si, SiO_2, Si_3N_4, etc. Lateral mask erosion and faceting, which occur during sputter etching, result in sloped wall profiles and patterns whose dimensions do not faithfully replicate those of the mask. In addition, ion bombardment of photoresist, or polymers in general, can cause the resist to degrade (carbonize) from the scission of chemical bonds such as C–H, C–N and C–O.

In order to avoid these problems in inert ion etching, particularly when etching deep, higher-aspect ratio features, a small partial pressure of O_2 was added to the inert working gas, e.g., Ar, and a thin film of reactive metal such as Ti, Al, or Cr used in place of photoresist as the etch mask (Cantagrel, 1975, Cantagrel and Marchal, 1973). The thin metallic mask is patterned using photoresist and wet chemical etching or by the lift-off technique. The etch rate of the metallic mask in Ar/O_2 is significantly reduced from the value in pure Ar, since the surface oxide layers produced (TiO_2, Cr_2O_3, etc.) have greatly reduced sputtering yields. For example, the etch rate of Ti in Ar^+ (500 eV, 1 mA/cm^2) is reduced from about 340 to 45 Å/min when O_2 is introduced at partial pressures of $\geqslant 2 \times 10^{-5}$ Torr

(Robertson, 1978). The etching rate of the substrate material is assumed to be unchanged or only slightly affected by the slight (5–15%) O_2 addition. The addition of a reactive gas such as O_2 to an inert ion beam etching system to increase mask-to-substrate selectivity is probably the earliest commercial exploitation of RIBE (c. 1973) (Cantagrel and Marchal, 1973).

At low partial pressures of oxygen ($\lesssim 5 \times 10^{-5}$ Torr), the use of Ar/O_2 mixtures in Kaufman-type ion sources will not significantly reduce hot-filament lifetimes. When larger oxygen concentrations are used or when O_2^+ ion beam etching is performed, hot filaments should be replaced with hollow cathode or other electron emitters less susceptible to oxidizing ambients. For example, when running pure oxygen in a Kaufman-type source, hot-filament cathodes reportedly burned out within only two hours (Pak and Sites, 1979).

In contrast to materials such as Ti, Al, and Cr, whose ion beam etch rates are very low in partial pressures of oxygen, organic polymers such as photoresist and polyimide have etch rates under O_2^+ ion bombardment which are very high. For example, under Ar^+ ion bombardment (700 eV, 0.7 mA/cm^2), it has been reported that the erosion rates of Al and polyimide are comparable – 600 and 450Å/min, respectively (Harper et al., 1982). On the other hand, under O_2^+ bombardment (700 eV, 0.7 mA/cm^2), the Al etch rate decreases to only ~70 Å/min, while the polyimide rate increases to about 2700 Å/min (Harper et al., 1982). The etch rate increased only weakly with O_2^+ energy in the range 300–1000 eV. RIBE etching of polyimide in O_2^+ is therefore rapid and highly selective (40:1) with respect to Al. Other workers have also reported that etching of polymers in ion beams derived from oxygen proceeds much more rapidly than in inert ion beams. For example, DeGraff and Flanders (1979) measured rates of approximately 720 Å/min for polyimide under 500 eV, 0.5 mA/cm^2 O_2^+ bombardment. They found in addition that O_2^+ RIBE etching was highly anisotropic and free of redeposition, enabling them to etch grating patterns having line widths as narrow as 400 Å in 2000 Å thick polyimide. The mask used in these studies was a thin 400 Å film of Cr patterned using PMMA resist and lift-off technique. Gokan et al. (1984) have also studied O_2^+ beam etching of resist materials and achieved high selectivity with respect to metals (e.g., >50:1 for AZ1350J over gold at 100 eV). In addition, they find the resist etch rate depends strongly on oxygen pressure – a fact attributed to ion-assisted neutral oxygen molecules.

In analogy with ashing of photoresist in an O_2 plasma, etching of polymers in O_2^+ beams is presumably enhanced by the chemical reactivity of the substrate (C_xH_y) to O_2. When volatile product molecules (CO, CO_2,

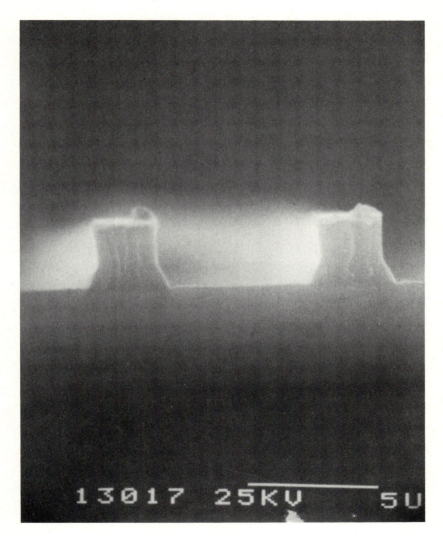

Fig. 47. Thin Al layer over photoresist over Si, etched in oxygen ion beam. The Al was patterned before RIBE of the underlying two layers. (Harper et al., 1982.)

H_2O) are formed, the etch rate is enhanced. As Harper et al. (1982) have pointed out, however, the erosion rate of polymers in O_2^+ beams (yields of ~ 100 Å/incident ion) are much too large to be accounted for simply by counting the number of reactive oxygen atoms available in the incident beam, and some additional mechanism(s) must be in operation such as radiation or ion-enhanced activation of adsorbed oxygen to account for the 'extra' material actually removed. When nonvolatile or low sputter rate surface films are produced (Al_2O_3, TiO_2), the rate is suppressed. Figure 47 shows an etch profile in photoresist obtained by Harper et al. using O_2^+ ion beam etching (Harper et al., 1982). A thin reactive metal film (evaporated Al patterned using a thin resist and Ar^+ etching) was employed to mask the thick photoresist layer during RIBE with O_2^+. The edge profile in the resist is seen to be highly anisotropic, while the profile in the underlying Si displays a sloped sidewall and some trenching. Trilevel resist schemes (thin resist, thin metal mask, thick planarizing resist) such as this one are of increasing interest in VLSI applications, primarily to allow devices with high steps or irregular surface topography to be patterned with high resolution.

As a variation on the etching of photoresist in O_2^+ beams, Castellano (1983) has studied the etching of photoresist (AZ 1350J) under Ar^+ bombardment with O_2 leaked into the etch chamber. Oxygen was backfilled so as to comprise 25–100% of the total 8×10^{-5} Torr system operating pressure. Figure 48 shows the etch rates of resist, polySi and SiO_2 under 500 eV

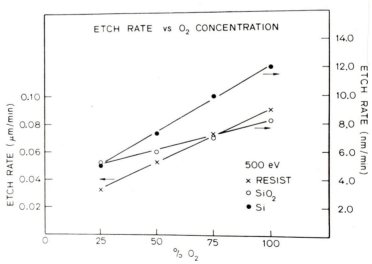

Fig. 48. Etch rates of photoresist, polysilicon and SiO_2 under Ar^+ bombardment (500 eV, 0.5 mA/cm^2) versus O_2 partial pressure in the etch chamber (Castellano, 1983).

Ar^+ bombardment (0.5 mA/cm^2) in the presence of various partial pressures of O_2. Resist erosion rates are seen to be a factor of ~10 greater than that of SiO_2 or polySi. The absolute resist etch rates with 100% O_2 in the chamber (~850 Å/min) are comparable to those obtained on polymers under O_2^+ bombardment (500 eV, 0.5 mA/cm^2). Profiles under Ar^+/O_2 were reported to be highly anisotropic.

As might be expected, carbonaceous materials such as vitreous carbon and graphite also exhibit etch rate enhancement in O_2^+ beams. For example, carbon has one of the lowest Ar^+ sputter rates known (~20 Å/min at 500 eV, 0.5 mA/cm^2), but under O_2^+ bombardment, this rate increases to about 1000 Å/min (Kaufman, 1978).

A particularly novel use of RIBE with O_2^+ was reported by Adesida et al. (1982). In this work, a finely-focused high-energy Si^+ ion beam (40 keV, 1×10^{15} cm^{-2}) was used to 'write' a pattern onto the resist surface by ion implantation. Under subsequent O_2^+ bombardment, the Si-implanted areas rapidly formed SiO_2 at the surface, which acted as a local etch mask protecting the underlying resist from O_2^+ RIBE. As a result, a high-resolution pattern was directly transferred into the resist layer.

RIBE of polymeric materials other than resist can also be effectively carried out in Ar/O_2 mixtures. Figure 49 shows a pattern of 15 μm deep × 5 μm wide square sections etched in polyester (Robertson, 1978). Because of ion-bombardment-induced surface crosslinking, inert ion beam milling alone is not effective. Instead, a gas such as O_2 is added to prevent crosslinking by scavenging free radical sites as they are formed on the polymer surface. Actually, the pattern shown in fig. 49 was RIBE etched at ≃ 2000 Å/min using a three-gas $Ar/O_2/He$ admixture, the He being used to increase cooling of the heat-sensitive polymer. An aluminum mask was used which, under the conditions used, etched 15 times slower than the polymer. Additional RIBE etch results for thinner polyimide layers (~1.5 μm) are shown in fig. 50 (Courtesy of Commonwealth Scientific Corp.). In this case, the polyimide was spun over sputtered Al and patterned with both an inorganic (SiO_2) and a photoresist mask. Ion beam energies of approximately 600–900 eV and current densities of 0.5–2 mA/cm^2 gave polyimide etch rates of 200–2500 Å/min. The SEM photographs (fig. 50) show that vertical wall profiles can be obtained. It was reported that this RIBE etch was highly selective with respect to the underlying Al (30–40 : 1) and that slight overetching effectively cleared the bottom of the via holes of residual contamination that could degrade the electrical conductivity of the contact.

Fig. 49. 5 µm × 5 µm square patterns etched 15 µm deep into polyimide by RIBE using Ar/O$_2$/He admixture (Robertson, 1978; courtesy of D. Bollinger; reprinted with permission from Solid State Technology, Technical Publishing, a company of Dun & Bradstreet).

Fig. 50. RIBE etch of polyimide layer (~1.5 µm thick) over aluminum. A bilevel SiO$_2$/photoresist mask was used. (Anonymous 1982; Courtesy of Commonwealth Scientific Corp.)

We now turn our attention to the effect of RIBE on organic masking materials when gases other than O_2 or Ar/O_2 are employed. RIBE etch rates of polymeric resist materials have been reported for a variety of gases, including CCl_4 (Downey et al., 1981), C_2F_5 (Heath, 1981a, 1982), CF_4 (Matsui et al., 1980), CF_4/O_2 (Horiike et al., 1979; Matsui et al., 1981), C_4F_8 (Matsuo and Adachi, 1982), $SiCl_4$ (Matsuo and Adachi, 1982), and CHF_3 (Meusemann, 1979). In general, different resist materials (e.g., AZ1350J, AZ1470, HPR-204, PMMA, SEL-N, COP) exhibit similar RIBE etch rates which are typically in the range of about 200–600 Å/min. Etch rate ratios of SiO_2 : resist are typically in the range 2–5. A notable exception in this regard is the work of Horiike et al. (1979), in which CF_4/O_2 admixtures (50% O_2) gave resist etch rates of 1500 Å/min, which were 15 times that of SiO_2 (Baudrant et al., 1983). The work of Matsuo and Adachi (1982) is also interesting in that ion beams extracted from $SiCl_4$ etched SiO_2 *slower* than either AZ 1350J or PMMA (Heath, 1982). Using C_4F_8, however, the oxide etch rate was 3–5 times higher than that of the resist.

Although the above resist etch rates apply to RIBE performed with the incident beam normally incident on the substrate, work of Matsui et al. has shown that the incidence angle of the reactive ion beam does *not* appear to have a pronounced effect on resist etch rate (Matsui et al., 1980). For example, using CF_4 (500 eV, 0.4 mA/cm^2), the RIBE rate of AZ 1350J was nearly constant for incidence angles of 0–50°, decreasing by only about 10%. This behavior was quite different from that observed with inert ion bombardment. Under Ar^+ bombardment (500 eV, 0.4 mA/cm^2), the etch rate increased by about 50% over the same angular range.

RIBE is generally thought of as an anisotropic technique. However, in cases where the resist is intentionally tapered, RIBE can produce controlled sidewall angles. If the etch rate of the resist and substrate are not too different, the slope of the tapered resist will be more-or-less faithfully transferred into the substrate. If the substrate etches more rapidly than the resist, a more vertical taper will be transferred into the substrate. Figure 35 (section 5.2.1) shows a result of etching 3750 Å of n-type polysilicon over thermal SiO_2. The polysilicon was masked with 1.1 µm of HPR-204 positive resist, which had been intentionally tapered. Under the conditions of the RIBE etch (CCl_4, 750 eV, 0.8 mA/cm^2), the relative etch rates of polysilicon : resist are about 2 : 1. Figure 35 shows that the sloped photoresist has left the polysilicon sidewall with an even taper and that the sidewall is at a steeper angle than that made by the resist, as expected from the etch rate ratio.

Under typical RIBE conditions (500 eV, ~0.5–1 mA/cm^2), about 0.25–0.5 W/cm^2 is incident on the sample surface. Excessive temperature rise ($T \approx 150°C$) can cause many resists to degrade and flow, giving rise to non-faithful pattern transfer and making stripping of resist difficult. The reader is referred to Egerton et al. (1982) for a further discussion of these effects and ways of preventing them. In general, some type of heat sinking must be used to prevent the resist from thermally degrading or flowing. Vacuum grease, gallium, thermally-conductive silicones such as Thermalcote™ and MUNG™, and electrostatic hold-downs have all been used to hold samples to a cooled platen. Methods involving adhesives, of course, are unwieldy and can introduce contamination. A more promising approach is to physically clamp the samples to a cooled platen and flow a thermally conductive gas (e.g., Ar, He) between the wafer back surface and the platen. The 'gas-assisted cooling' can be very effective and is the technique used on the RIBE system (Varian Model RE-580) in our laboratory.

5.4. New applications

As with other dry etch techniques, RIBE has been most often applied to patterning those semiconductors, dielectrics and metals which are of current interest in IC device fabrication. These include single-crystal and polycrystal Si, SiO$_2$ PSG, Si$_3$N$_4$ and Al-based metallizations (Al, Al–Si [~2%], Al–Si [~2%]–Cu [1–4%]) masked with conventional photoresists or e-beam resists. In this section, we consider several applications of RIBE which are relatively new, but promising.

5.4.1. Refractory metals and metal silicides

One such application is the use of RIBE to pattern refractory metal silicides such as MoSi$_2$ (Powell, 1983) and TaSi$_2$ (Bollinger, 1983; Baudrant et al., 1983). It is generally agreed that doped polysilicon is not the optimal material for gate electrodes and interconnects in VLSI devices. Its limiting high resistivity (~750–1000 μΩ cm) leads to excessive RC time-constant delays as channel length and gate electrode width are scaled down below the 1 μm level. As a replacement for polysilicon, metal silicides have attracted much attention – used either by themselves or together with polysilicon in a so-called 'polycide' stack (metal silicide/polysilicon) over the gate oxide. For example, MoSi$_2$ has a sheet resistance about 10 times lower than heavily doped polysilicon and can withstand high-temperature oxidizing ambients.

RIBE of $MoSi_2$ films using reactive ion beams extracted from a CCl_4 plasma has recently been reported (Powell, 1983). The samples were thin films (~2500 Å thick) deposited onto (100) Si substrates by co-sputtering from elemental Si and Mo targets. Figure 51 presents RIBE etching rates of as-deposited $MoSi_2$ films using CCl_4 at selected beam energies from 500–1000 eV. The RIBE etch rate appears to increase linearly with ion current density.

Figure 52 compares the etch rates of $MoSi_2$ obtained using RIBE and inert ion beam etching (Ar^+) from 0.4–1 keV (current density = 0.35 mA/cm^2). The removal rate of $MoSi_2$ under RIBE is seen to be suppressed relative to the Ar^+ sputtering rate. At 500 eV, it is lower by ~40%. It was suggested that this may be the result of a protective coating such as carbon which accumulated on the sample surface and impeded the formation of

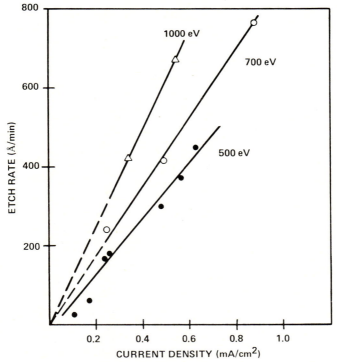

Fig. 51. RIBE etch rate of co-sputtered $MoSi_2$ film in CCl_4 for selected beam energies (Powell, 1983).

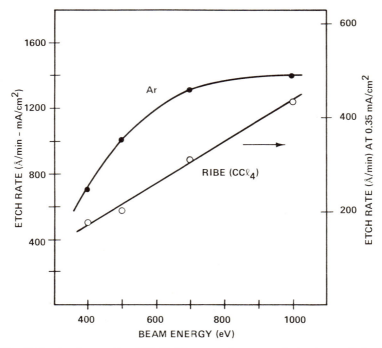

Fig. 52. Etch rate of co-sputtered $MoSi_2$ film versus ion energy for ion beams derived from argon and CCl_4 (0.35 mA/cm^2) (Powell, 1983).

volatile reaction products (Powell, 1983). Figure 53 shows the etch profile obtained after RIBE etching (500 eV, 0.6 mA/cm^2, CCl_4) through a 2500 Å $MoSi_2$ film and about 1 µm into the underlying Si substrate. The photoresist was stripped prior to electron microscopy. The etch profile displays good anisotropy (aspect ratio of about 5 : 1) with minimal trenching and no undercutting of the Si under the silicide. The relative etch rates of 1350J photoresist, $MoSi_2$ and Si were ~1 : 1 : 2.

In using RIBE to pattern polycide structures, whether based on $MoSi_2$ or some other likely silicide such as WSi_2, $TaSi_2$, $TiSi_2$, one wants to anisotropically etch both silicide and polysilicon layers with high selectivity relative to a thin (~200–400 Å) underlying gate oxide. Figure 54 shows a simplified cross-section of a polycide gate in an NMOS device about to be dry etched. In order to quantitatively estimate the relative etch rates of poly-Si, silicide, and oxide required to dry etch this structure, we follow

Fig. 53. Edge profile of RIBE-etched 0.25 μm $MoSi_2$ film over Si (CCl_4, 500 eV, 0.6 mA/cm^2) with photoresist removed (Powell, 1983).

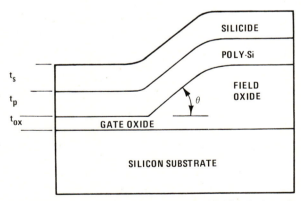

Fig. 54. Simplified cross-section of polycide gate in an NMOS device prior to dry etching (after Nang et al., 1982).

the approach presented in Nang et al. (1982) and make the following assumptions. We assume that the polysilicon and silicide films (thicknesses t_p, t_s) conformally coat the sloped transition region of the adjacent, thick, field oxide. The slope of this region is taken to be $\theta \sim 30°$. Nonuniformity of film thickness and RIBE etch are $F = \pm 5\%$ and $E = \pm 2\%$, respectively. Finally, we assume that one can tolerate $\alpha = 80\%$ of the gate oxide's thickness (t_{ox}) being etched above the source and drain regions.

Successful etching requires that the silicide/polysilicon stack in the sloped region be cleared without overetching the gate oxide. A simple calculation (Nang et al., 1982) allows one to estimate the required etch rate ratio of polysilicon-to-oxide (R_p/R_o) in terms of the polysilicon-to-silicide etch rate ratio (R_p/R_s). Namely,

$$R_p/R_o \gtrsim \frac{[t_p + t_s(R_p/R_s)]}{\alpha t_{ox}} \left[\frac{(1+F)}{(1-E)\cos\theta} - \frac{(1-F)}{(1+E)} \right]. \quad (5)$$

Using the assumed values of θ, E, F, and α for a 1500 Å silicide/1500 Å polysilicon gate on a $t_{ox} = 300$ Å gate oxide, we get

$$\frac{R_p}{R_o} \gtrsim 1.9 \left[1 + \frac{R_p}{R_s} \right]. \quad (6)$$

For the case of RIBE etching $MoSi_2$ with CCl_4 at 500 eV, $R_p/R_s \sim 0.5$, so that a polysilicon-to-oxide etch ratio of 3:1 is required – several times larger than was in fact observed. If a sufficiently selective, reactive gas chemistry can be developed, however, RIBE offers the attractive possibility of a one-step polycide etching process.

Figure 55 shows the result of RIBE etching a $TaSi_2$/polysilicon structure over SiO_2 (Baudrant et al., 1983). The $TaSi_2$ and polysilicon are each 2000 Å thick. Beam energy was 600 eV, current density = 0.7 mA/cm^2, and the beam was incident at 15° to the sample normal. A 10% overetch was used. Using Ar^+, the etch rates of $TaSi_2$, polysilicon and SiO_2 were all about 325 Å/min. With either pure Cl_2 or $Cl_2(50\%)/Ar(50\%)$ gas mixtures in the ion source, the etch rates of $TaSi_2$ and polysilicon increased to about 1000 Å/min, while that of SiO_2 remained the same. Selectivity with respect to the underlying gate oxide was thus about 3:1. Good anisotropy was obtained (aspect ratio of approximately 4:1) and with a slight 10% overetch, no residue was left at the SiO_2 surface. The use of CCl_4 as an etch gas, however, did leave a residue on the underlying SiO_2. The etch rate of the LPCVD polysilicon did not depend on how it was doped, whether by ion implantation, diffusion, or in-situ doping during LPCVD growth.

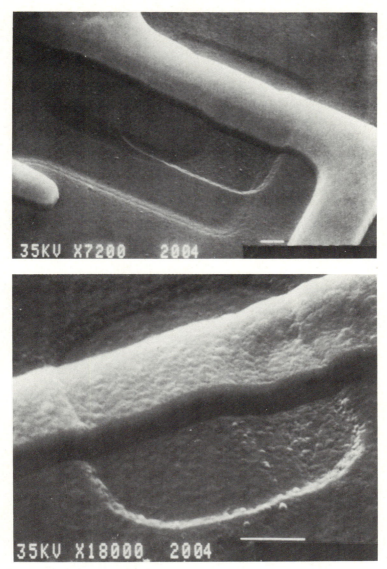

Fig. 55. RIBE etch of $TaSi_2$/polysilicon bilayer over SiO_2 using Cl_2, 600 eV, 0.7 mA/cm². Resist has been removed. (Baudrant et al., 1983; reprinted with permission from Solid State Technology, Technical Publishing, a company of Dun & Bradstreet).

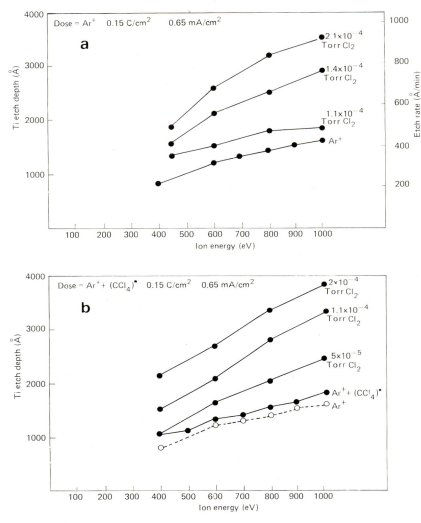

Fig. 56. Etch rate of Ti under bombardment with ions extracted from (a) Ar or (b) CCl$_4$ plasma as a function of ion energy for various partial pressures of Cl$_2$ in the chamber (Chinn et al., 1983a).

In addition to silicides of refractory metals, the pure refractory metals themselves are of interest in IC processing. For example, NMOS self-aligned silicide process ('salicide process') uses films of Mo or Ti deposited over both polysilicon and oxide (Okabayashi et al., 1982; Lau et al., 1982). RIBE etching of Mo and Ti have both been reported.

Chinn et al. (1983a) have used the technique of ion-beam-assisted etching to etch the refractory metals Mo and Ti – with quite different results. In the case of Ti (see fig. 56), the etch rate was found to be strongly dependent on the flux of molecular and atomic chlorine. This was true for both inert-ion-beam (Ar^+) and reactive-ion-beam (CCl_3^+) assisted etching in a chlorine ambient. The Ti etch rate increased with the flux of chlorine delivered either by the incident ion beam (increased ion current density) or from collisions with the background chlorinated gases (increased background gas partial pressure). In the case of Mo, however, neither increased partial pressures of chlorine in the etch chamber nor increased reactive, chlorinated ion beam flux density had a significant effect on etch rate. Presumably, the difficulty of removing low vapor pressure molychlorides limits the etching rate of Mo.

The use of fluorinated ion beams is expected to improve this situation for Mo, since MoF_6 is highly volatile. For Ti, low volatility of potential chlorides is not a concern; however, the presence of small partial pressures of O_2/H_2O can significantly lower the ion-beam-assisted etch rates through formation of TiO_2 (low etch rate).

As discussed earlier, the use of thin Ti films as a dry etch mask for inert ion-beam milling was made possible by the addition of small partial pressure of O_2 to the etch chamber. In the present case, however, we want to pattern the Ti so that the presence of O_2 or H_2O is highly undesirable. Also undesirable is H_2, which by scavenging reactive Cl through HCl formation, can reduce the etch rate.

5.4.2. *Carbides*

Silicon carbide (SiC) is an unconventional silicon compound which has recently been patterned with RIBE (Matsui et al., 1981). SiC is a semiconductor which, because of its heat resistance and mechanical/chemical stability, has been proposed for use in solid-state devices operating at high temperature (> 500°C), in high levels of radiation, etc. Matsui et al. report fabricating SiC grating patterns (1 μm period) with RIBE, using $CF_4 + O_2$ gas (Matsui et al., 1981). SiC etch rates in CF_4 and CF_4/O_2 (40%) (500 eV, 0.4 mA/cm^2) were 65 Å/min and 130 Å/min, respectively. The most durable etch mask appeared to be Cr, which had an etch rate of only 30 Å/min in CF_4/O_2 (40%).

5.4.3. III–V compound semiconductors

In the context of Si device fabrication, pattern transfer by dry etching is now an established process technology. By comparison, relatively little has been reported on dry etching of III–V compound semiconductors, although activity in this area is clearly increasing. In Gbit electronics, for example, the use of dry etching to fabricate short channel ($< 0.5\,\mu m$) GaAs MESFETs is attractive. In integrated optics as well, dry etching could be used to advantage to fabricate such devices as light-emitting diodes, distributed-feedback lasers, optical gratings, and waveguides. Dry etching of III–V materials using RIBE has been carried out by:

(1) Introducing reactive gases such as Cl_2 or CCl_4 into a Kaufman-type source, and

(2) Ion-beam-assisted etching, whereby the sample is bombarded by an Ar^+ ion beam in the presence of a high partial pressure of Cl_2.

Chlorine-based chemistries are usually chosen for these RIBE and ion-assisted etch applications, since the potential chlorides of Group III and Group V elements are volatile to some extent.

By introducing Cl_2 gas directly into a Kaufman source, Bosch et al. (1981) have carried out RIBE of InP. As noted in section 1, an important feature of RIBE is the ability, in principle, to control edge profile by varying the incidence angle of the impinging reactive ions. The study of Bosch et al. demonstrates how this worked out in practice for the InP/Cl_2 system. Figure 57 shows the etch rate and etched sidewall inclination from the normal as a function of ion beam incidence angle θ. Ion current density is $0.6\,mA/cm^2$. A thin Ti metal film ($\sim 2000\,\text{Å}$) patterned by lift-off technique, was used as an etch mask since the etch rate of photoresist by chlorine was unacceptably high. Note that although the RIBE etch rate of InP increases strongly with reactive-ion energy (the rate at 1000 eV is ~ 4 times that at 500 eV), there is little dependence on incidence angle for θ in the range 0–50°. On the other hand, the inclination of the etched sidewall changed by over 30°.

Under normal incidence RIBE, the sidewalls slope outward by about 17°. Etching with the sample at $\theta \approx 24°$ to the beam, however, produced a mesa in which one sidewall was vertical, while the opposite wall was inclined at $\sim 30°$. In order to obtain vertical sidewalls on *both* sides of the mesa, the sample was inclined at an angle of about 42° to the beam and

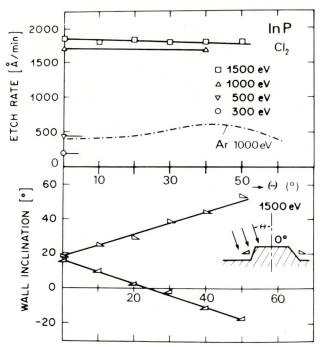

Fig. 57. Etch rate and sidewall angle produced in InP under RIBE with Cl_2 as a function of beam incidence angle (Bosch et al., 1981).

rotated about the sample normal at 2 rpm. The SEM photograph in fig. 58 shows the result of such a RIBE etch of InP. The mesa displays vertical sidewalls, although the bottom portion tapers away from the edge – an effect attributed to shadowing during the spinning process. This taper could be virtually eliminated by using incline angles of $\lesssim 20°$ during the etch, although in this case etch anisotropy was reduced (aspect ratio $\lesssim$ 6:1).

Yuba et al. (1983) have also reported RIBE of InP with Cl_2 as well as with CCl_2F_2 and CHF_3, although the highest etch rate (~0.2 μm/min) was obtained for Cl_2 (0.3 mA/cm^2, $E \gtrsim 1.5$ keV). For example, at 1 keV the etch rate with Cl_2 (1600 Å/min) is nearly an order of magnitude larger than with CCl_2F_2. Rates with CCl_2F_2 and CHF_3 were actually lower than with inert Ar^+ ion milling.

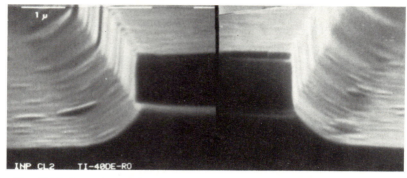

Fig. 58. SEM photomicrograph of InP after RIBE etch in Cl_2 where sample was angled 45° with respect to ion beam and rotated at about 2 rpm (Bosch et al., 1981).

RIBE etching of GaAs has been reported by Powell for CCl_4 introduced into a Kaufman source (Powell 1982). Etch rates (normalized to 1 mA/cm^2 ion current density) are shown in fig. 59 and compared with inert ion milling (Ar$^+$) at ion energies in the range 400–800 eV. Both RIBE and Ar$^+$-sputter etch rates vary linearly with beam energy, although the RIBE etch rate is about twice as great as purely physical sputtering. Etch rates of ~5800 Å/min · mA/cm^2 were obtained at 750 eV. In addition, good selectivity over conventional 1350J photoresist (~12:1) allowed deep features to be etched without the need for metal masks. The data in fig. 59 was obtained with the sample located about 22 cm from the ion source extraction grids. Keeping incident ion flux density constant, a reduction in source-to-sample distance by ~30%, *increased* the etch rate by the same percentage. Under these conditions, GaAs etch rates of 7800 Å/min · mA/cm^2 could be obtained at 750 eV. The Ar$^+$ sputter rates, normalized to incident ion flux, were independent of source-to-sample distance. Presumably, at distances closer to the source, a higher density of reactive free radicals (Cl atoms?) was encountered, which in turn increased the etch rate. These neutral species diffuse through the extraction grids and downstream into the etch chamber. A similar variation of etch rate with distance has been observed in RIBE of AlCuSi alloys using CCl_4 (Downey et al., 1981). Edge profiles under normal incidence reactive ion bombardment were anisotropic but not completely vertical. Instead, sidewalls sloped away from the original mask edge at an angle of about 16° to the ion beam. This same 'overcut' taper was observed for both $\langle 100 \rangle$- and $\langle 111 \rangle$-oriented GaAs substrates. As in the work of Bosch et al. (1981) on

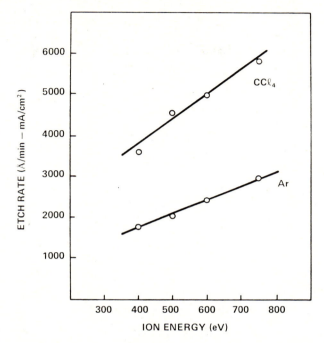

Fig. 59. Etch rate versus ion energy for GaAs etched in ion beams derived from Ar and CCl$_4$ (Powell, 1982).

InP, more vertical sidewalls could be obtained by performing RIBE at oblique incidence. A 10° tilt angle, for example, produced sidewalls inclined only 12° from the vertical.

Asakawa and Sugata (1983) have reported RIBE of GaAs and GaAlAs with Cl$^+$, Cl$_2^+$ ions extracted from an electron cyclotron resonance plasma (2.45 GHz). Their results showed that nearly equal (within about 10%) etch rates for GaAs and Ga$_{1-x}$Al$_x$As could be obtained for a wide range of ternary compound composition ($x = 0.15$–0.3), beam energy (400–500 eV), and gas pressure (10^{-4} to 2×10^{-3} Torr). In RIE, equal etch rates have been difficult to obtain due, in part, to the presence of hard-to-remove Al$_2$O$_3$ on the Ga$_x$Al$_{1-x}$As surface.

An unconventional approach to RIBE etching III–V materials has recently been presented by Geis et al. (1981). In this case, a chemically reactive gas beam from a jet and an inert ion beam from the Kaufman ion

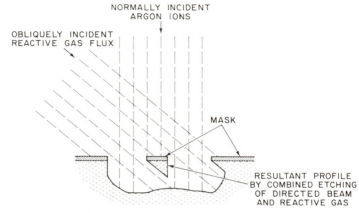

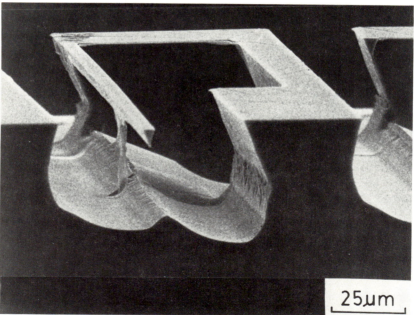

Fig. 60. Etching of GaAs under simultaneous Ar^+ ion beam and neutral Cl_2 (gas jet) bombardment can produce novel topographic structures (Geis et al., 1981).

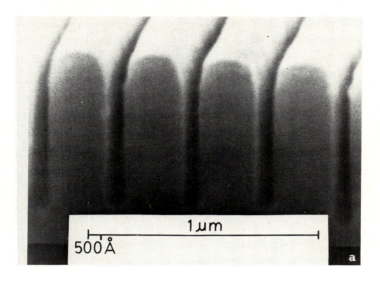

Fig. 61. Results of inert ion-beam-assisted etching of GaAs demonstrating ultra-high resolution when reactive Cl_2 flux and Ar^+ ion beam are normal to surface. (a) Structures are 500 Å wide slots etched 7500 Å deep (Geis et al., 1981); (b) structures are 1.5 μm deep slots (Lincoln et al., 1983).

source simultaneously impinge on the sample surface. The partial pressure of the reactive gas at the sample surface can be quite high (several tens of mTorr), and by adjusting the angle between gas jet and ion beam, a variety of novel surface topographies can be generated (fig. 60). Using a Cl_2 gas jet located about 1–2 mm above the edge of the sample (reactive gas flux $\sim 5 \times 10^{18}\,\text{cm}^{-2}\,\text{s}^{-1}$), and Ar^+ ion beam (500 eV, 1 mA/cm^2), etch rates for GaAs of 3–5 μm/min could be obtained. With both jet and ion beam normal to the sample surface, highly anisotropic etching ($\sim 50:1$ aspect ratio) was obtained. Figure 61a shows an SEM of such an etched GaAs sample. The structure consists of 500 Å wide slots etched ~ 0.75 μm deep. Figure 61b

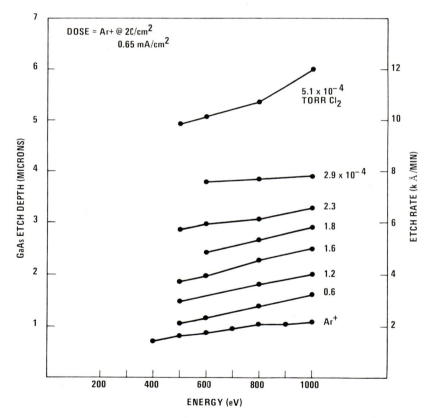

Fig. 62. Etch rate of GaAs under Ar^+ bombardment (0.4–1 keV, 0.65 mA/cm^2) in the presence of various partial pressures of Cl_2 (Chinn et al., 1983a).

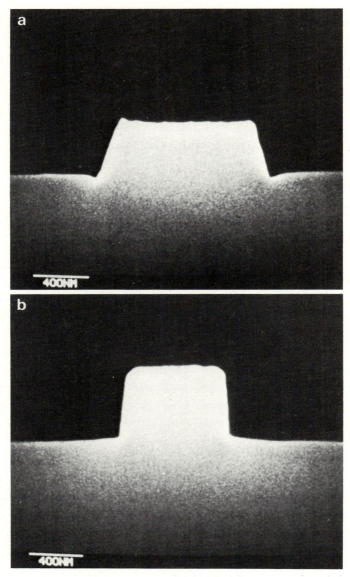

Fig. 63. SEM micrographs showing how sidewall profile can be tailored in inert ion-beam-assisted etching of GaAs. (a) Ar^+, 0.8 keV, 0.65 mA/cm^2; (b) Ar^+, 0.5 keV, 0.65 mA/cm^2 with 3×10^{-4} Torr chlorine. Etch depth is ~5100 Å in each case. (Chinn et al., 1983a.)

shows more recent results of 1.5 µm deep slots etched in GaAs by Lincoln et al. (1983) using a similar technique. Although this technique is very promising, it may be difficult to maintain a high degree of uniformity and directionality, and over large-diameter samples. Of course, a directed gas jet is not the only way to increase the partial pressure of reactive species at the sample. An alternative is to simply backfill the etch chamber with the desired gas, although as a practical matter, pumping large amounts of gases such as Cl_2 is unattractive. Chinn et al. (1983a) have in fact reported etching GaAs by backfilling the etch chamber with up to about 2×10^{-4} Torr of Cl_2 and then performing Ar^+ bombardment. This enabled them to achieve etch uniformity better than ±5% over a GaAs wafer.

Figure 62 shows the GaAs etch rates obtained by Chinn et al. (1983a) under 0.4–1 keV Ar^+ bombardment (0.65 mA/cm^2) in partial pressures of Cl_2 as high as 5×10^{-4} Torr. It is clear that the etch rate depends on both reactive gas pressure and, to a lesser extent, Ar^+ ion energy. Under 700 eV Ar^+ bombardment, a removal rate of about 2000 Å/min is obtained in the absence of Cl_2. This rate has increased to nearly 10 000 Å/min in the presence of 5×10^{-4} Torr of Cl_2. In addition, at increased Cl_2 concentrations, the wall profiles were altered from an overcut (positive) slope to nearly vertical, as shown in fig. 63. A similar situation was observed when using chlorinated reactive ion beams (CCl_4), although in this case, a smaller increase in Cl_2 partial pressure was required to obtain vertical sidewalls.

6. RIBE-induced damage

During RIBE, the substrate is subjected to bombardment by energetic ions (~0.1–1 keV) and UV photons originating in the ion source. In general, bombardment by both reactive and inert ions must be considered a possibility. For example, when a reactive gas such as CF_4 is introduced directly into the ion source, reactive species such as $C_xF_y^+$ are produced. In addition, if a hollow-cathode electron source were used to produce the CF_4 discharge, some component of the extracted ion beam will be inert – Ar^+, Xe^+, etc. – due to gas flowing through the hollow cathode and into the ion source. On the other hand, in ion-beam-assisted etching, the incident ion beam is comprised mainly of inert ions, whereas the partial pressure of reactive gas at the substrate is high. Reactive gas molecules which then diffuse to the ion source extraction grids can be ionized and contribute a reactive ion component to the incident ion beam.

The topic of damage induced by dry processing (and how to avoid or anneal out such damage) is of practical concern in semiconductor device fabrication. Patterning of SiO_2/Si is of special interest since electron or hole traps created by damage in the oxide can produce undesired shifts in the threshold voltage of MOSFETs. Device damage can be caused in a variety of ways: by atom displacements during sputtering, and knock-on mixing, creation of electron–hole pairs in field and gate oxides by energetic particles or photons, surface contamination by redeposition or polymer formation, contamination by incorporation of impurities originating in the ion source (carbon), from neutralizer filaments (heavy metals) or from the etch gases themselves (Cl_2, F_2, H_2), and so on. Selected references (Hosaka et al., 1981; Deppe et al., 1977; Yabumoto et al., 1981; Pang et al., 1982; Ephrath and DiMaria, 1981; Matsumoto and Sugano, 1982; Singh et al., 1983) discuss device and material damage by inert ion beam etching and RIE of Si and SiO_2 layers.

With regard to RIBE, early work by Meusemann (1979) reported that little damage was caused by RIBE (1 keV, $0.8\,mA/cm^2$) of SiO_2 using CF_3 or CF_4/O_2 (4%). RIBE was first used to remove the thick field oxide over the active area of an MOS device and a thin dry oxide regrown. The Si/SiO_2 interface properties were then determined by the high-frequency C–V technique. Fixed oxide charge and fast interface state densities were equal to those obtained after wet chemical etching. More recently, workers have studied RIBE-induced damage of Si and SiO_2 using ion beams extracted from a C_2F_6 discharge (800 eV, $0.4\,mA/cm^2$) (Gildenblat and Heath, 1981). In these studies, 1000 Å of thermal SiO_2 over Si(100) was etched back 600 Å by either wet chemical etching (buffered HF) or RIBE. MOS capacitors were then formed on the remaining 400 Å of SiO_2 by evaporation and patterning of 0.5 μm thick Al contacts, followed by a 30 min, 400°C H_2 sinter. High frequency (1 MHz) and quasi-static C–V measurements showed that negligible damage was introduced in the oxide as a result of RIBE. On the other hand, if the initial 1000 Å oxide was regrown, the resulting MOS capacitors displayed C–V plots with significant (~0.8 V) flatband potential shifts. It was estimated that slow interface states of average density ~$1.6 \times 10^{12}\,cm^{-2}\,eV$ were created by RIBE bombardment of the Si(100) surface. In addition, a buildup of F at the SiO_2/Si interface was observed – presumably the result of the fluorocarbon ion bombardment.

In general, more work needs to be done to understand the nature of the chemical and physical damage caused by RIBE and to investigate ways of avoiding such damage. One possibility, for example, would be to work at lower ion beam energies to minimize the undesired effects of physical sputtering when a damage-sensitive material or interface is being etched. Beam energy can have a dramatic effect on the creation of damage. For example, the number of interface states created at the SiO_2/Si interface of an oxidized, Ar^+-bombarded Si sample decreased by ~10 when beam energy was reduced from 1 keV to 0.25 keV (Pang et al., 1982). Although relatively little has been done on RIBE-induced damage *per se*, increasing attention is being paid to the damage caused by other dry etch techniques (RIE, planar plasma) and by directed beams of energetic particles such as focused electrons and ions for fine-line lithography applications. Knowledge gained in these areas will in turn be of value in understanding the damage mechanisms of reactive ion beams.

7. Conclusion

RIBE and Ion-Assisted Chemical Etching offer advantages for submicron etching of material and multicomponent systems such as AlCuSi, metal silicides and polycides, and III–V materials (GaAs, InP, AlGaAs), where a delicate balance between the chemical and physical components of the

Table 5
Equipment developments and research topics.

Equipment developments
1. Continued development of reactive-gas-tolerant microwave ion sources.
2. Wider use of hollow cathodes and plasma bridge neutralizers.
3. Refinement of existing electrical end-point detection schemes (Downey et al., 1981) and/or development of other process monitoring techniques.
4. Development of robust, single-extraction grids for very low-energy RIBE.

Research topics
1. Directional O_2^+ etching of polymers.
2. Etching very smooth, high aspect ratio patterns in III–V materials for optoelectronic applications.
3. RIBE-induced damage.
4. One-step etching of polycide structures.
5. Continued basic studies of ion-assisted reactive gas-surface chemistry.
6. Increased selectivity by new gas chemistries or control of ionic species leaving the ion source.
7. Dry etching with directional, reactive *neutral* atoms having high energy (~100–1000 eV).

etching is required to achieve optimal results. Improvements in selectivity and throughput should make these techniques commercially viable for these applications. Table 5 suggests topics where increased research and development efforts could advance RIBE technology.

The balance between the physical and chemical components of etching is provided by the independent control over the critical process parameters. This allows for precise tailoring of a process for individual requirements. The choice of which parameters are varied and which mode of RIBE is used should be carefully selected in consideration of the specific system being dry etched. It must be noted that it is important to achieve a correct balance between the chemical and physical component. If there is too much of the chemical component, there can be problems with residues (e.g., AlCuSi), undercutting; or if ion nonvolatiles are created (inappropriate choice of chemistry), a decreased etch rate (e.g., $MoSi_2$ in CCl_4). If, on the other hand, there is too much of the physical component, there are problems with overcutting, trenching, redeposition, low selectivity, and lower etch rates. When these components are well matched, however, RIBE and Ion-Assisted Chemical Etching can produce the high-resolution, anisotropic etching required for future VLSI device requirements.

References

Adesida, I., J.D. Chinn, L. Rathbun and E.D. Wolf, 1982, J. Vac. Sci. Technol. **21**, 666.
Anonymous, 1981, Solid State Technol., October 1981, p. 100.
Anonymous, 1982, Reactive Ion Beam Etching (RIBE) of Polyimides, Application Note #206-82 (Commonwealth Scientific Corp., Alexandria, VA).
Asakawa, K., and S. Sugata, 1983, Japan. J. Appl. Phys. Lett. **22**, L653.
Barker, R.A., T.M. Mayer and W.C. Pearson, 1982, Proc. Electrochem. Soc. Spring Meeting, Vol. **82-1**, Extended Abstract 276, p. 452.
Barker, R.A., T.M. Mayer and W.C. Pearson, 1983, J. Vac. Sci. Technol., **B1**, 37.
Baudrant, A., A. Passerat and D. Bollinger, 1983, Solid State Technol., September 1983, p. 183.
Bollinger, L.D., 1983, Solid State Technol., January 1983, p. 99.
Bollinger, D., and R. Fink, 1980, Solid State Technol., December 1980, p. 97.
Bosch, M.A., L.A. Coldren and E. Good, 1981, Appl. Phys. Lett. **38**, 264.
Brown, D.M., B.A. Heath, T. Coutumas and G.R. Thompson, 1980, Appl. Phys. Lett. **37**, 159.
Burggraaf, P.S., 1980, Semicond. Int., January 1980, p. 61.
Cantagrel, M., 1975, J. Vac. Sci. Technol. **12**, 1340.
Cantagrel, M., and M. Marchal, 1973, J. Mater. Sci. **8**, 1711.
Castellano, R.N., 1983, A Comparison of Reactive Ion Beam Milling and Reactive Ion

Etching for Multilevel Resist Patterning, in: Proc. Int. Ion Eng. Congress, Kyoto, Japan, p. 1527.

Chapman, B., and M. Nowack, 1980, Semicond. Int., November 1980, p. 139.

Chinn, J.D., A. Fernandez, I. Adesida and E.D. Wolf, 1983a, J. Vac. Sci. Technol. **A1**, 701.

Chinn, J.D., I. Adesida and E.D. Wolf, 1983b, Appl. Phys. Lett. **43**, 185.

Coburn, J.W., and H.F. Winters, 1979, J. Appl. Phys. **50**, 3189.

DeGraff, P.D. and D.C. Flanders, 1979, J. Vac. Sci. Technol. **16**, 1906.

Deppe, H.R., B. Hasler and J. Kopfner, 1977, Solid State Electron. **20**, 51.

D'Heurle, F.M., and P.S. Ho, 1978, Electromigration in Thin Films, in: Thin Films Interdiffusion and Reactions, eds. J.M. Poate, K.N. Tu and J.W. Mayer (Wiley, New York) p. 243.

Downey, D.F., and R.A. Powell, 1983, A Review of Reactive Ion Beam and Ion Assisted Chemical Etching, in: Plasma Processing, eds. G.S. Mathad, G.C. Schwartz and G. Smolinsky (The Electrochemical Society, Pennington, NJ).

Downey, D.F., W.R. Bottoms and P.R. Hanley, 1981, Solid State Technol. February 1981, p. 121.

Dzioba, S., G. Este, R.A. Bond and H.M. Naguib, 1981, Proc. Electrochem. Soc. Fall Meeting, Vol. **81-2**, Extended Abstract 259, p. 628.

Egerton, E.J., A. Nef, W. Millikin, W. Cook and D. Baril, 1982, Solid State Technol., August 1982, p. 84.

Ephrath, L.P., and D.J. DiMaria, 1981, Solid State Technol., April 1981, p. 182.

Geis, M.W., G.A. Lincoln, N. Efremow and W.J. Piacentini, 1981, J. Vac. Sci. Technol. **19**, 1390.

Gerlach-Meyer, U., 1981, Surf. Sci. **103**, 524.

Gerlach-Meyer, U., J.W. Coburn and E. Kay, 1980, J. Appl. Phys. **51**, 3362.

Gildenblat, G., and B.A. Heath, 1981, Proc. Electrochem. Soc. Fall Meeting, Vol. **81-2**, Extended Abstract 284, p. 693.

Gokan, H., M. Itoh and S. Esho, 1984, J. Vac. Sci. Technol. **B2**, 34.

Hakhu, J.K., 1981, Reactive Ion Beam Etching for VLSI, in: Proc. Symp. on VLSI Technol., Hawaii, 1981, (IEEE, New York) Paper 5-2, p. 66.

Haring, R.A., A. Haring, F.W. Saris and A.E. deVries, 1982, Appl. Phys. Lett. **41**, 174.

Harper, J.M.E., J.J. Cuomo, P.A. Leary and G.M.S. Summa, 1981, J. Electrochem. Soc. **128**, 1077.

Harper, J.M.E., J.J. Cuomo and H.R. Kaufman, 1982, J. Vac. Sci. Technol. **21**, 737.

Heath, B.A., 1981a, Proc. Electrochem. Soc. Fall Meeting, Vol. **81-2**, Extended Abstract 283, p. 690.

Heath, B.A., 1981b, Solid State Technol., October 1981, p. 75.

Heath, B.A., 1982, J. Electrochem. Soc. **129**, 1396.

Horiike, Y., M. Shibagaki and K. Kadono, 1979, Japan J. Appl. Phys. **18**, 2309.

Hosaka, S., Y. Kawamoto and S. Hashimoto, 1981, J. Vac. Sci. Technol. **18**, 17.

Jolly, T.W., and R. Clampitt, 1982, Ion Milling – The Competing Technology, in: Proc. Semicond. Int., Birmington, UK, 1982 (Cahners, UK).

Kaufman, H.R., 1974, Technology of Electron-Bombardment Ion Thrusters, in: Adv. Electronics and Electron Phys., Vol. 36, ed. L. Marton (Academic, New York), p. 265.

Kaufman, H.R., 1978, J. Vac. Sci. Technol. **15**, 272.

Kaufman, H.R., J.J. Cuomo and J.M.E. Harper, 1982a, J. Vac. Sci. Technol. **21**, 725.

Kaufman, H.R., J.M.E. Harper and J.J. Cuomo, 1982b, J. Vac. Sci. Technol. **21**, 764.
Lau, C.K., Y.C. See, D.B. Scott, J.M. Bridges, S.M. Perma and R.D. Daview, 1982, Titanium Disilicide Self-Aligned Source/Drain + Gate Technology, in: Proc. Int. Electron Device Meeting (IEEE, New York) p. 714.
Lincoln, G.A., M.W. Geis, S. Pang and N.N. Efremow, 1983, J. Vac. Sci. Technol. **B1**, 1043.
Maddox, R.L., and M.R. Splinter, 1980, Dry Processing Methods, in: Fine Line Lithography, Vol. 1, ed. R. Newman (North-Holland, Amsterdam) ch. 4.
Matsui, S., T. Yamato, H. Aritome and S. Namba, 1980, Japan. J. Appl. Phys. Lett. **19**, L126.
Matsui, S., S. Mizuki, T. Yamato, H. Aritome and S. Namba, 1981, Japan. J. Appl. Phys. Lett. **20**, L38.
Matsumoto, H., and T. Sugano, 1982, J. Electrochem. Soc. **129**, 2823.
Matsuo, S., 1983, Reactive Sputter Etching and Reactive Ion Beam Etching, in: Proc. Int. Ion Eng. Congress, Kyoto, Japan, p. 1597.
Matsuo, S., and Y. Adachi, 1982, Japan. J. Appl. Phys. **21**, L4.
Mayer, T.M., and R.A. Barker, 1981, Proc. Electrochem. Soc. Fall Meeting, Vol. **81-2**, Extended Abstract 282, p. 688.
Mayer, T.M., and R.A. Barker, 1982a, J. Electrochem. Soc. **129**, 585.
Mayer, T.M., and R.A. Barker, 1982b, J. Vac. Sci. Technol. **21**, 757.
Mayer, T.M., R.A. Barker and L.J. Whitman, 1981, J. Vac. Sci. Technol. **18**, 349.
Merchant, P.P., 1982, Hewlett–Packard Journal, August 1982, p. 28.
Meusemann, B., 1979, J. Vac. Sci. Technol. **16**, 1886.
Miyake, K., S. Tachi, K. Yagi and T. Tokuyama, 1982, J. Appl. Phys. **53**, 3214.
Miyamura, M., O. Tsukakoshi and S. Komiya, 1982, J. Vac. Sci. Technol. **20**, 986.
Miyamura, M., T. Uchiyama, O. Tsukakoshi and S. Komiya, 1983, Reactive Ion Beam Etching of Si and SiO_2 with an Electron Cyclotron Resonance Ion Source, in: Proc. Int. Ion Eng. Congress, Kyoto, Japan, p. 1623.
Nang, K.L., T.C. Holloway, R.F. Pinizzotto, Z.P. Sobczak, W.R. Hunter and A.F. Tasch, Jr., 1982, IEEE Trans. Electron Dev. **ED-29**, 547.
Okabayashi, H., E. Nagasawa and M. Morimoto, 1982, Low Resistance MOS Technology Using Self-Aligned Refractory Silicidation, in: Proc. Int. Electron Dev. Meeting (IEEE, New York) p. 556.
Okano, H., and Y. Horiike, 1980, Proc. Electrochem. Soc. Spring Meeting, Vol. **80-1**, Extended Abstract 111, p. 291.
Okano, H., and Y. Horiike, 1982, Japan. J. Appl. Phys. **21**, 696.
Pak, S-J., and J.R. Sites, 1979, Broad Beam Ion Source Operation with Four Common Gases, in: ONR Tech. Report SF24 (Dept. of Phys., Colorado State Univ., Ft. Collins, CO, September 1979).
Pang, S., D.D. Rathman, D.J. Silversmith, R.W. Mountain and P.D. DeGraff, 1982, Proc. Electrochem. Soc. Fall Meeting, Vol. **82-2**, Extended Abstract 184, p. 389.
Powell, R.A., 1982, Japan. J. Appl. Phys. Lett. **21**, L170.
Powell, R.A., 1983, J. Electrochem. Soc. **130**, 1164.
Reader, P.D., D.P. White and G.C. Isaacson, 1978, J. Vac. Sci. Technol. **15**, 1093.
Robertson, D.D., 1978, Solid State Technol., Dec. 1978, p. 57.
Rovell, P.J., and G.F. Goldspink, 1981, J. Vac. Sci. Technol. **19**, 1398.
Sakaki, H., Y. Sekiguchi and K. Yokoyama, 1981, J. Vac. Sci. Technol. **19**, 23.

Schaible, P.M., and G.C. Schwartz, 1979, J. Vac. Sci. Technol. **16**, 337.
Singh, R., S.J. Fonash, S. Ashok, P.J. Caplan, J. Shappiro, M. Hage-Ali and J. Ponpon, 1983, J. Vac. Sci. Technol. **A1**, 334.
Speidell, J.L., J.M.E. Harper, J.J. Cuomo, A.W. Kleinsasser, H.R. Kaufman and A.H. Tittle, 1982, J. Vac. Sci. Technol. **21**, 824.
Stein, R.J., 1982, Microelectronic Manufacturing and Testing, August 1982, p. 22.
Tachi, S., K. Miyake and T. Tokuyama, 1982, Japan. J. Appl. Phys., Suppl. **21-1**, 141.
Thompson Jr., G.R., 1981, Reactive Ion Beam Etching with the Millatron G. Ion Source, in: Proc. Amer. Vac. Soc. S. Cal. Meeting (Anaheim, CA) p. 11.
Tokunaga, K., and D. Hess, 1980, J. Electrochem. Soc. **127**, 928.
Tolliver, D.L., 1980, Solid State Technol., November 1980, p. 99.
Tsukada, T., 1983, J. Electron. Eng., September 1983, p. 79.
Tu, Y-Y., T.J. Chuang and H.F. Winters, 1981, Phys. Rev. **B23**, 823.
Watakabe, Y., S. Matsuda, T. Kato and H. Nakata, 1983, Reactive Ion Beam Etching of Aluminum Alloys, in: Proc. Int. Ion Eng. Congress, Kyoto, Japan, p. 1607.
Weiss, A., 1982, Semicond. Int., October 1982, p. 69.
Winters, H.R., 1983, private communication.
Winters, H.R., J.W. Coburn and T.J. Chuang, 1983, J. Vac. Sci. Technol. **B1**, 469.
Yabumoto, N., M. Oshima, O. Michikami and S. Yoshii, 1981, Japan. J. Appl. Phys. **20**, 893.
Yuba, Y., K. Gamo, X.G. Ho, Y.S. Zhang and S. Namba, 1983, Japan. J. Appl. Phys. **22**, 1211.

CHAPTER 5

DRY ETCHING FOR MICROELECTRONICS – A BIBLIOGRAPHY

L.C. MOLIERI

IBM
General Products Division
San Jose, California 95193, USA

Dry Etching for Microelectronics, edited by R.A. Powell
© *Elsevier Science Publishers B.V., 1984*

Contents

1. Introduction 217
2. Organization 217
3. Sources of information 220
    3.1 Conventional 220
    3.2. Unconventional 221
4. Bibliography 223
    4.1 General: Books, reviews, conference proceedings and general interest . . . 223
    4.2. Plasma etching 228
    4.3. Reactive ion etching 254
    4.4. Ion beam milling and reactive ion beam etching 272
    4.5. End-point detection and plasma diagnostics 283
    4.6. Resists and masking materials 288

1. Introduction

At the annual meeting of the Electrochemical Society in 1976, Alan Reinberg reviewed the selective removal of material by reaction with a chemically reactive glow discharge (A.R. Reinberg, 'Plasma Etching in Semiconductor Manufacture – A Review', in: *Etching for Pattern Definition*, Electrochem. Soc., 1976, pp. 91–110). He noted that, 'Although a few articles have appeared and several recent meetings have included talks on the subject of plasma etching, the major source of information is still equipment manufacturers' literature, patents and informal discussions'. He went on to add that, 'One gets the impression on reading much of this literature that the author is often trying to sell the process, and that dispassionate scientific articles are few'. Whether today's scientific papers on plasma etching are more dispassionate is a matter of conjecture; however, the number of such papers has grown phenomenally. By 1970, only a handful of papers had been published on dry etching, while by 1980 the yearly publication rate had increased to over 130 (fig. 1). This trend has continued, and today there are numerous scientific and technical papers dealing with the subject of dry etching. The present chapter consists of over 800 references to these published works, making it one of the most extensive bibliographies on dry etching available.

2. Organization

In this bibliography, articles dealing with the use of plasma techniques for purposes other than etching or pattern transfer have not been cited. Thus the extensive literature on plasma-enhanced chemical vapor deposition (PECVD) is not included. Readers with an interest in this field are referred to the work of D.T. Hawkins (*Chemical Vapor Deposition, 1960–1980, A Bibliography*, IFI Plenum Publishing Co., New York, 1981). Also excluded is the large body of patent literature relevant to dry etching; however, patents may be located online through any of several commercial data base systems, or in the U.S. Patent Office *Official Gazette*, which is

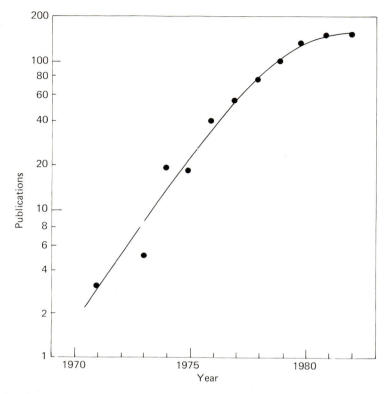

Fig. 1. Growth in the yearly number of papers published on dry etching from 1970–1982. Patents or the use of plasma in deposition (e.g. plasma-enhanced chemical vapor deposition) have not been included.

held in most university and many larger public libraries. In addition, useful citations which fall outside the scope of this bibliography or were inadvertently missed may be found in the extensive references following each of the critical review chapters in the present volume.

This bibliography was compiled electronically, using the online search facility of DIALOG Information Services, Inc., of Palo Alto, California. Using a microcomputer, modem and telephone, the DIALOG data bases were searched and printouts of relevant citations obtained. Data bases which were searched correspond to the following abstracting sources: Chemical Abstracts, Computer and Control Abstracts, Dissertation

Abstracts International, Electrical and Electronics Abstracts, Engineering Index, Government Reports Announcements, Metals Abstracts, Physics Abstracts, Science Citation Index, Surface Coatings Abstracts, and World Aluminum Abstracts. Most of these data bases index literature from the late 1960's or early 1970's to the present; since plasma technology is relatively new, we feel that a high percentage of the relevant literature is included here. Literature published prior to the late 1960's may be retrieved from the appropriate printed abstracting sources.

A core group of keywords was matched against each data base. These included:

Dry Etching; End-Point Detection; Ion Beam Etching; Ion Milling; Plasma Etching; Plasma Processing; Reactive Ion Beam Etching (RIBE); Reactive Ion Etching (RIE); Reactive Sputter Etching; Resists and Masking Processes/Materials.

Because of the large number of citations (over 1000 retrieved, of which over 800 appear in this bibliography), it was decided to group references by general subject rather than list them in one long sequence. Citations which fell into more than one subject (e.g., a paper describing end-point detection for RIE) were placed in the group deemed primary to the author's research. Therefore, none of the citations appears in more than one category. Subject groups were chosen to represent logical divisions within the overall group of plasma processing technologies:

(1) General (books, general reviews, bibliographies, etc.);
(2) Plasma Etching;
(3) Reactive Ion Etching;
(4) Ion Beam Milling and Reactive Ion Beam Etching;
(5) End-Point Detection and Plasma Diagnostics;
(6) Resists and Masking Materials.

Within each subject category, citations are arranged alphabetically by first author and subdivided alphabetically by joint authors, and, finally, by titles. References without a personal author appear at the beginning of each category under the heading 'Anonymous'.

Every attempt has been made to follow a standard format for citations. Journal articles, which comprise the largest group of references, appear in the following general format:

⟨Author(s)⟩.
⟨Title of Article⟩.
⟨Journal Name⟩ ⟨Volume (Issue)⟩: ⟨Page Numbers⟩, ⟨Date⟩.

Other references (conference papers, technical reports, government documents, etc.) follow the format of the original citation. References to conference papers include full information on the location and date of the conference.

In addition to obtaining an abstract or full text of the references cited here, the reader may wish to addend this work by compiling additional, updated references. Firms of medium to large size maintain technical libraries equipped to provide online literature searching, document delivery, interlibrary loan service, and referral to local information sources. Newer and smaller firms frequently cannot provide extensive library service, leaving the researcher to rely on personal or outside sources. Fortunately, numerous information resources and services are available to researchers on a regular or 'as needed' basis. The major data base vendors offer updating services (frequently known as SDI – Selective Dissemination of Information). These services are based on a 'profile' of the researcher's subject interests. As new publications are added to the data bases, items which match the profile are printed and mailed. In addition, an increasing number of firms offer research and document delivery service. They generally have access to major libraries (some have representatives at research libraries nation-wide) and can provide copies of documents in a matter of days, depending on the source. These companies can generally be found in the telephone company Yellow Pages under 'Library Research and Services'. In addition to providing copies of documents, many of them will perform literature searches on demand. The major data base vendors frequently have contracts with a number of document delivery firms, and offer online ordering facilities, transmitting orders electronically to the vendor of your choice. For more information on this type of service, consult your library or any of the document delivery firms.

3. Sources of information

3.1. Conventional

Although peer-reviewed papers on dry etching for microelectronics regularly appear in a large number of journals and periodicals, the majority have appeared in only a dozen or so – principally: Journal of the Electrochemical Society, Proceedings of the Electrochemical Society Meeting, Journal of Applied Physics, Applied Physics Letters, Japanese Journal of

Applied Physics, Japanese Journal of Applied Physics Letters, IEEE Transactions on Electron Devices, IEEE Electron Device Letters, Journal of Vacuum Science and Technology, Electronics Letters, Thin Solid Films, Solid State Technology, and Semiconductor International. In addition, conference proceedings, selected chapters in books on general semiconductor processing, and monographs devoted to plasma etching (such as the present one) are available today. A wealth of information can readily be retrieved from articles found in these sources and by following up references cited therein. A useful recent bibliography appears in: Landolt–Bornstein. Numerical Data and Functional Relationships in Science and Technology. New Series, Group 3, Volume 17c: *Technology of Si, Ge and SiC*, eds. K.-H. Hellwege and O. Madelung (Springer-Verlag, New York, 1984) p. 328–366.

3.2. Unconventional

A variety of less conventional sources of information on dry etching is also available, which we now list and illustrate with selected examples.

(1) *Trade Magazines.* Periodicals such as *Solid State Technology* and *Semiconductor International* regularly contain contributed articles on dry etching, as well as up-to-date equipment reviews by staff writers. Particularly valuable are topical issues devoted to plasma processing, such as Solid State Technology, April 1983.

(2) *Vendors' Literature.* Dry etching accounts for over $100 million in yearly semiconductor equipment sales, representing over 25 major equipment vendors. Most of these companies maintain customer demonstration/applications laboratories where personnel will discuss plasma processing issues (at least with regard to results obtained on their own equipment). Often, useful information is available from the manufacturer which has not been published elsewhere. Lists of vendors with their addresses/phone numbers can be obtained from directories published yearly by the magazines *Circuits Manufacturing* (Vendor Directory Issue) and *Semiconductor International* (Buyer's Guide) or by contacting SEMI, the Semiconductor Equipment and Materials Institute (Mountain View, CA). A particularly useful publication is the Microelectronics Manufacturing and Testing 1983 Desk Manual (Lake Publishing, Libertyville, IL) which not only lists suppliers of dry etching products but includes definitions of technical terms and brief summary articles on relevant topics.

(3) *Privately-Sponsored Plasma Seminars.* Complementing the symposia held by scientific associations, there are several yearly seminars on dry etching sponsored by equipment manufacturers which present invited and

sometimes contributed papers. The proceedings can usually be purchased even if one does not attend the actual conference. Corporate sponsors include Tegal (10th Annual Plasma Seminar to be held May 21, 1984, Palo Alto, CA) and ET Electrotech (5th Advanced Plasma Technology Seminar to be held March 27–29, 1984, Monterey, CA).

(4)   *Trade Shows.* The technical sessions accompanying yearly microelectronic-oriented trade shows such as SEMICON/West, its spin-offs (SEMICON/East, Southwest, Europe and Japan), and the Advanced Semiconductor and Equipment Exposition (ASEE, San Jose, CA) often include invited talks or topical sessions on dry etching. In most cases, proceedings are available after the show to nonattendees for a price.

(5)   *Short Courses.* A variety of short courses on plasma etching, lasting from a day to a week, are offered by private companies, professional societies, and educational institutions. Recent examples in these categories include:

- (a) 'Plasma Processing for Microelectronic Fabrication: Plasma Deposition, Etching and Sputtering of Thin Films for VLSI', H.H. Swain and R. Reif, Summer Session of the Massachusetts Institute of Technology, July 18–22, 1984, Cambridge, MA.
- (b) 'Plasma Chemistry for IC Fabrication', D.W. Hess and A.T. Bell, May 8, 1983 (held in conjunction with the 163rd meeting of the Electrochemical Society, San Francisco, CA).
- (c) 'Plasma Etching', J. Coburn, a one-day Short Course of the American Vacuum Society (available as 'Plasma Etching and Reactive Ion Etching', J.W. Coburn, AVS Monograph Series, monograph editor N. Rey Whetten).
- (d) 'Plasma Etching', a three-day course taught by S. Broydo et al., October 19–21, 1983, Palo Alto, CA (Continuing Education Institute, Los Angeles, CA).
- (e) A three-part video-taped short course entitled 'Plasma Processing for VLSI Fabrication (Part I: Plasma Fundamentals; Part II: VLSI Deposition; Part III: VLSI Etching)' from the Engineering Renewal and Growth Program, Colorado State University, Fort Collins, CO.

Probably the easiest way to find out in advance about these and other short courses is to consult the time schedules for coming events published regularly in the magazines *Semiconductor International* (Calendar of Events), *Solid State Technology* (Calendar) and *Physics Today* (Calendar), or society newsletters such as those of The American Vacuum Society.

4. Bibliography

4.1. General: Books, reviews, conference proceedings and general interest

Anonymous.
Development and industrialization of gas plasma etching technology (LSI processing).
Research and Development in Japan Awarded the Okochi Memorial Prize, 83–87, 1976.

Anonymous.
Plasma etching technology and applications, 1970–1981. (Citations from the Searchable Physics Information Notices data base.)
National Technical Information Service, Report Number PB82–859 208, Dec. 1981.

Anonymous.
Progress of VLSI submicron structures attained in laboratory.
Elektronikschau 58 (11): 38–40, Nov. 11982.

Anonymous.
Research at Siemens: Think today for tomorrow.
Funkschau No. 5: 71–74, 4 March 1983.

Arai, E.
VLSI fine technology and its problems.
Japanese Journal of Applied Physics, Part I 21 (Suppl. 21-1): 43–49, 1982.

Ballantyne, J.M.
Microfabrication techniques for submicron devices.
Microcircuit Engineering 82. International Conference on Microlithography, Grenoble, France, 5–8 Oct. 1982, p. 35–43.

Baron, M.; Zelez, J.
Choosing a vacuum system for plasma and vapour deposition processes.
Semiconductor Production 16, 19, 21, 25, 27, Autumn 1980.

Baron, M.; Zelez, J.
Vacuum systems for plasma etching, plasma deposition, and low pressure CVD.
Solid State Technology 21 (12): 61–65, 1978.

Bell, A.T.
An introduction to plasma processing.
Solid State Technology 21 (4): 89–94, Apr. 1978.

Bersin, R.L.
Programmed plasma processing: The next generation.
INTERFACE '78, Proceedings of the Microelectronics Seminar, San Diego, California, 1–3 Oct. 1978, p. 21–28.

Bersin, R.L.
Using low-temperature plasmas for surface treatment of polymers.
Polymer News 2 (11–12): 13–18, 1976.

Bersin, R.L.; Gelernt, B.
Faithful pattern transfer: What are the limits in production?
SPIE Proceedings 334: 163–174, 1982.

Bursky, D.
Chipping away the barriers to the million-device IC.
Electronic Design 30 (16): 91–98, Aug. 5, 1982.

Chang, C.W.; Szekeley, J.
Plasma applications in metals processing.
Journal of Metals 34 (2): 57–64, Feb. 1982.

Chapman, B.
Glow discharge processes. Sputtering and plasma etching. (Book).
New York: John Wiley, 1980.

Clark, H.A.
Plasma processing at moderate vacuum.
Solid State Technology 19 (6): 51–54, 1976.

Crabtree, P.N.; Gorin, G.; Thomas, R.S.
Plasma etching techniques in semiconductor manufacture – a review.
Scanning Electron Microscopy Conference, Los Angeles, California, 17–21 Apr. 1978, p. 543–544.

Dieleman, J. (ed.)
Proceedings, Symposium on Plasma Processing, 3rd, Denver, Colorado, 11–16 Oct. 1981.

Down, M.G.
Plasma processing for materials production.
Electric Power Research Institute, Palo Alto, California. Report Number EPRI EM 2771, Dec. 1982.

Dundas, P.H.
Plasma processing in Russia – a review.
Abstracts of Papers of the American Chemical Society, 1974, p. 38.

Duval, P.
Pumping chlorinated gases in plasma etching.
Journal of Vacuum Science and Technology A 2, Pt. 1): 233–236, Apr.–June 1983.

Egerton, E.J.; Nef, A.; Millikin, W.; Cook, W.; Baril, D.
Positive wafer temperature control to increase dry etch throughput and yield.
Solid State Technology 25 (8): 84–87, Aug. 1982.

Eklund, M.H.
Wafer processing in the early 1980's.
Semiconductor International 6 (1): 38–45, Jan. 1983.

Ephrath, L.M.
Dry etching for VLSI – a review.
Journal of the Electrochemical Society 128 (3): PC97, 1981.

Ephrath, L.M.
Dry etching for VLSI – a review.
Journal of the Electrochemical Society 129 (3): 62–66, Mar. 1982.

Ephrath, L.M.
Etching needs for VLSI.
Solid State Technology 25 (7): 87–92, July 1982.

Ephrath, L.M.; DiMaria, D.J.
Review of RIE induced radiation damage in silicon dioxide.
Solid State Technology 24 (4): 182–188, Apr. 1981.

Fujioka, T.
Dry etching challenges fine processing in the submicron technology.
JEE 19 (192): 60–64, Dec. 1982.

Hammond, V.J.; Norgate, P.
The case for ion beam etching.
Electronic Equipment News 13 (6): 48–56, Oct. 1971.

Hassan, J.K.; Sarkary, H.G.
Lithography for VLSI: An overview.
Solid State Technology 25 (5): 49–54, May 1982.

Hawkins, D.T.
Ion milling (ion-beam etching), 1954–1975: A bibliography.
Journal of Vacuum Science and Technology 12 (6): 1389–1398, Nov.–Dec. 1976.

Hawkins, D.T.
Ion milling (ion-beam etching), 1975–1978: A bibliography.
Journal of Vacuum Science and Technology 16 (4): 1051–1071, July–Aug. 1979.

Hutt, M.
Etching technology.
Sputtering and Plasma Etching Conference, 24th Sputtering School, Pebble Beach, California, 10–12 Dec. 1979.

Jacob, A.
Plasma processing – an art or a science?
Solid State Technology 26 (4): 151–155, Apr. 1983.

Jones, N.
Advanced sputtering and plasma etching required for VLSI.
Sputtering and Plasma Etching Conference, 24th Sputtering School, Pebble Beach, California, 10–12 Dec. 1979.

Kurogi, Y.
Recent trends in dry etching.
Thin Solid Films 92 (1–2): 33–40, 11 Juni 1982.

Lehmann, H.W.
Dry etching techniques.
European Conference on Electrotechnics, 4th: From Electronics to Microelectronics (Eurocon '80), Stuttgart, Germany, 24–28 March 1980, p. 57–62.

Maddox, R.L.
New microelectronic processing technology: A review of the state of the art.
Microelectronics Journal 11 (1): 4–18, 1980.

Marcoux, P.J.
Dry etching: An overview.
Hewlett–Packard Journal 33 (8): 19–23, Aug. 1982.

McCaughan, D.V.
The use of plasmas in CCD processing.
International Conference on Technology and Applications of Charge Coupled Devices, Edinburgh, Scotland, 25–27 Sept. 1974, p. 281–289.

Merz, J.L.
Monolithic integration of optical sources and detectors.
SPIE Proceedings 239: 53–60, 1980.

Mogab, C.J.
Ion beam, plasma and reactive ion etching.
European Solid State Device Research Conference, 9th, Munich, Germany, 10–14 Sept. 1979, p. 37–54.

Norgate, P.; Hammond, V.J.
Ion beam etching.
Physics and Technology 5 (3): 186–203, 1974.

Parry, P.D.; Rodde, A.F.
Anisotropic plasma etching of semiconductor materials.
Solid State Technology 22 (4): 125–132, Apr. 1979.

Petvai, S.I.; Schnitzel, R.H.
Some promising applications of ion milling in surface cleaning.
Surface Contamination: Genesis, Detection and Control, vol. 1.
Washington, D.C., 10–14 Sept. 1978, p. 297–311.

Poulsen, R.G.
Plasma etching in integrated circuit manufacture – a review.
Journal of Vacuum Science and Technology 14 (1): 266–274, Jan.–Feb. 1977.

Smith, H.I.
Fabrication techniques for optical and acoustical micro-electronic devices.
Symposium on Optical and Acoustical Micro-Electronics, New York, New York, 16–18 Apr. 1975, p. 221–232.

Somekh, S.
Reactive sputter etching and its applications.
SPIE Proceedings 334: 175–180, 1982.

Somekh, S.; Casey, H.C. Jr.
Dry processing of high resolution and high aspect ratio structures in $GaAs-Al_x-Ga_{1-x}-As$ for integrated optics.
Applied Optics 16 (1): 126–136, Jan. 1977.

Stach, J.; Woytek, A.J.
Dry etching systems expand IC device processing arsenal.
Industrial Research/Development 24 (7): 107–108, 110, July 1982.

Symposium on Advances in Ion Technology, Murray Hill, New Jersey, 26 May 1976. Sponsored by the American Vacuum Society.
Journal of the American Vacuum Society 13 (5), Sept.–Oct. 1976.

Sze, S.M.
VLSI technology overviews and trends.
Japanese Journal of Applied Physics Suppl.: 3–10, 1982.

Tagaki, T. (ed.).
Symposium on Ion Sources and Ion-Assisted Technology and International Workshop on Ion-Based Techniques for Film Formation, 5th, Tokyo and Kyoto, Japan, 1–5 June, 1981.

Ting, C.H.; Neureuther, A.R.
Applications of profile simulation for thin film deposition and etching processes.
Solid State Technology 25 (2): 115–123, Feb. 1982.

Tolliver, D.L.
Plasma processing in microelectronics – past, present, and future.
Solid State Technology 23 (11): 99–105, Nov. 1980.

Tsukada, T.; Ukai, K.
Reactive ion etching.
Vacuum Society of Japan Journal 23 (9): 415–424, 1980.

Tuck, J.
Plasma etching.
Circuits Manufacturing 22 (7): 69–77, July 1982.

Valles, J.A.; Diaz, G.G.
Plasma and microelectronics.
Mundo Electronico No. 110: p. 133–1137, Sept. 1981.

Vossen, J.L.
The preparation of substrates for film deposition using glow discharge techniques.
Journal of Physics E: Scientific Instruments 12 (3): 159–167, Mar. 1979.

Wang, D.N.K.; Maydan, D.
Dry etching technology for fine line devices (VLSI fabrication).
Solid State Technology 24 (5): 121–125, May 1981.

Weiss, A.
Plasma etching of aluminum: Review of process and equipment technology.
Semiconductor International 5 (10): 10 p. between p. 69 and 84, Oct. 1982.

Widmann, D.
Etching of small structures.
International Conference on Microlithography, Paris, France, 21–24 June 1977, p. 323–331.

Yasuda, Y.
Recent developments in dry processing for very large scale integration.
Thin Solid Films 90 (3): 259–270, Apr. 23, 1982.

4.2. Plasma etching

Anonymous.
Fully automatic aluminum plasma etching machine.
Japanese Industrial Technology Bulletin 8 (1): 15, Apr. 1980.

Anonymous.
Inline plasma chemistry aluminum etcher.
Solid State Technology 24 (12): 24–25, Dec. 1981.

Anonymous.
More accurate method monitors plasma etching.
Design News 34 (9): 17, 1978.

Anonymous.
Selective plasma etching.
Solid State Technology 19 (2): 10, 1976.

Abe, H.
A new undercutting phenomenon in plasma etching.
Japanese Journal of Applied Physics 14 (11): 1825–1826, Nov. 1975.

Abe, H.
Plasma etching overview: Expanding applications.
Circuits Manufacturing 18 (4): 22, 24, Apr. 1978.

Abe, H.
The application of gas plasma to the fabrication of MOS LSI.
Solid State Devices Conference, 6th, Tokyo, Japan, 2–3 Sept. 1974, p. 287–295.

Abe, H.; Nishioka, K.; Tamura, S.; Nishimoto, A.
Microfabrication of anti-reflective chromium mask by gas plasma.
Japanese Journal of Applied Physics 15 (Suppl. 15-1): 25–31, 1976.

Abe, H.; Sonobe, Y.; Enomoto, T.; Fukuwatari, H.; Komiya, H.
Etching characteristics of Si and Si compounds by gas plasma and its application to the fabrication of MOS LSI.
Journal of the Electrochemical Society 120 (3): 93C, Mar. 1973.

Adams, A.C.
Plasma planarization (IC processing).
Solid State Technology 24 (4): 178–181, Apr. 1981.

Adams, A.C.; Capio, C.D.
Edge profiles in the plasma etching of polycrystalline silicon.
Journal of the Electrochemical Society 128 (2): 366–370, Feb. 1981.

Ahn, K.Y.; Freedman, R.W.
Modification of magnetic properties via plasma etching.
IBM Technical Disclosure Bulletin 20 (9): 3717–3718, Feb. 1978.

Aitken, A.; Poulsen, R.G.; MacArthur, A.T.P.; White, J.J.
A fully plasma etched ion implanted CMOS process.
International Electron Devices Meeting, Technical Digest, Washington D.C., 6–8 Dec. 1976, p. 209–213.

Aktik, M.; Segui, Y.; Ai, B.
Growth of polymer films on compound semiconductors and dry etching process.
Journal of Applied Physics 50 (10): 6567–6569, 1979.

Alcorn, G.E.; Feeley, J.D.
Plasma etching via holes in sputtered quartz.
IBM Technical Disclosure Bulletin 17 (9): 2701–2702, Feb. 1975.

Anderson, R.E.
Post-gate plasma and sputter process effects on the radiation hardness of metal gate CMOS integrated circuits.
IEEE Transactions on Nuclear Science NS-25 (6): 1459–1464, Dec. 1978.

Arikado, T.; Horiike, Y.
Si and SiO_2 etching under low self-bias voltage.
Japanese Journal of Applied Physics Pt. 1, 22 (5): 799–802, May 1983.

Atanasova, E.D.; Kirov, K.I.; Pantchev, B.G.; Georgiev, S.S.
Some investigation of Si and SiO_2 surface etched in CF_4 or CF_4O_2 plasma.
Physica Status Solidi (A) Applied Research 59(2): 853–859, Jun. 1980.

Baker, M.A.
Plasma cleaning and the removal of carbon from metal surfaces.
Thin Solid Films 69 (3): 359–368, Jul. 1980.

Battey, J.F.
Design criteria for uniform reaction rates in an oxygen plasma (Si wafer etching).
IEEE Transactions on Electron Devices ED-24 (2): 140–146, Feb. 1977.

Beenakker, C.I.M.; VanDommelen, J.H.J.; Vandepoll, R.P.J.
Decomposition and product formation in CF_4–O_2 plasma-etching silicon in the afterglow.
Journal of Applied Physics 52 (1): 480–485, 1981.

Beinvogl, W.; Deppe, H.R.; Stokan, R.; Hasler, B.
Plasma-etching of polysilicon and Si_3N_4 in SF_6 with some impact on MOS device characteristics.
IEEE Transactions on Electron Devices ED-28 (11): 1332–1337, 1981.

Bell, G.
Plasma etching in fabrication of SIPMOS (Siemens Power MOSFETS).
Journal of the Electrochemical Society 127 (3): PC91, 1980.

Bennett, R.S.
Highly selective etching of SiO_2 using $CClF_3$ + H_2.
IBM Technical Disclosure Bulletin 25 (9): 4589, Feb. 1983.

Bennett, R.S.; Ephrath, L.M.
Selective and directional etching of polysilicon and WSi_2.
IBM Technical Disclosure Bulletin 25 (1): 33–34, June 1982.

Benzing, D.W.
Automated characterization techniques for plasma etching processes.
Microelectronics Measurement Technology Seminar, 3rd Annual, San Jose, California, 17–18 March 1981, p. 21–42.

Bergendahl, A.S.; Bergeron, S.; Harmon, D.L.
Optimization of plasma processing of silicon-gate FET manufacturing applications.
IBM Journal of Research and Development 26 (5): 580–589, Sept. 1982.

Bernacki, S.E.
Low-pressure anisotropic plasma etching of doped polysilicon in CCl_4.
Journal of the Electrochemical Society 129 (3): PC105, 1982.

Berndlmaier, E.; Hryckowian, J.N.
Polysilicon resistor integrated planar design.
IBM Technical Disclosure Bulletin 24 (11A): 5619–5621, Apr. 1982.

Bersin, R.L.
Chemically selective, anisotropic plasma etching (IC processing).
Solid State Technology 21 (4): 117–121, Apr. 1978.

Bersin, R.L.
Plasma etching of thin metal and dielectric films.
Journal of Vacuum Science and Technology 13 (1): 169, Jan.–Feb., 1976.

Bersin, R.L.
Survey of plasma-etching processes.
Solid State Technology 19 (5): 31–36, May 1976.

Bhattacharya, S.
System for varying the directionality in plasma etching.
IBM Technical Disclosure Bulletin 20 (3): 991, Aug. 1977.

Blake, H.H. Jr.; Gajda, J.J.
Junction delineation by plasma etching.
IBM Technical Disclosure Bulletin 24 (1B): 434–6, Jun. 1981.

Blakeslee, M.C.; Freedman, R.W.; Krongelb, S.; Romankiw, L.T.
Aluminum etch mask for plasma etching.
IBM Technical Disclosure Bulletin 21 (3): 1256–1258, Aug. 1978.

Bond, R.A.; Dzioba, S.; Naguib, H.M.
Temperature measurements of glass substrates during plasma etching.
Journal of Vacuum Science and Technology 18 (2): 335–338, Mar. 1981.

Bondur, J.A.; Clark, J.A.
Plasma etching for $SiO_2/2$ profile control.
Solid State Technology 23 (4): 122–128, Apr. 1980.

Bondur, J.A.; Frieser, R.G.
Shaping of profiles in SiO_2 by plasma etching.
Journal of the Electrochemical Society 127 (3): PC89, 1980.

Bonet, C.
Thermal plasma processing.
Chemical Engineering Progress 78 (12): 63–69, 1976.

Booth, R.C.; Heslop, C.J.
Application of plasma etching techniques to metal oxide semiconductor (MOS) processing.
Thin Solid Films 65 (1): 111–123, Jan. 1980.

Bower, D.H.
Planar plasma etching of polysilicon using using CCl_4 and NF_3.
Journal of the Electrochemical Society 129 (4): 795–799, 1982.

Boyd, H.; Tang, M.S.
Applications for silicon tetrafluoride in plasma etching.
Solid State Technology 22 (4): 133–138, Apr. 1979.

Braga, E.S.; Mammana, A.P.; Mammana, C.I.Z.; Anderson, R.L.
Plasma etching of SnO_2 films on silicon substrates.
Thin Solid Films 73 (2): L5–6, Nov. 1980.

Bresnock, F.J.; Stumpf, T.
Implementation of adaptive process-control to a dry etching process.
Journal of Vacuum Science and Technology 20 (4): 1027–1030, 1982.

Brown, H.L.; Bunyard, G.B.; Lin, K.C.
Applications of mass spectrometers to plasma process monitoring and control (for semiconductor device processing).
Solid State Technology 21 (7): 35–38, Jul. 1978.

Bruce, R.H.
Anisotropy control in dry etching.
Solid State Technology 24 (10): 64–68, Oct. 1981.

Bruce, R.H.; Reinberg, A.R.
Profile control with DC bias in plasma etching.
Journal of the Electrochemical Society 129 (2): 393–396, 1982.

Bunyard, G.B.; Raby, B.A.
Plasma process development and monitoring via mass spectrometry.
Solid State Technology 20 (12): 53–57, Dec. 1977.

Burggraaf, P.S.
Plasma etching technology.
Semiconductor International 2 (10): 54–58, Dec. 1979.

Burstell, C.B.; Hung, R.Y.; McMullin, P.G.
Preferential etch scheme for GaAs–GaAlAs.
IBM Technical Disclosure Bulletin 20 (6): 2451, Nov. 1977.

Burton, R.H.; Temkin, H.; Keramidas, V.G.
Plasma separation of InGaAsP/InP light-emitting diodes.
Applied Physics Letters 37 (4): 411–412, 15 Aug. 1980.

Busta, H.H.; Lajos, R.E.; Kiewit, D.A.
Control plasma etching.
Industrial Research and Development 20 (6): 5 p. between p. 133 and 142, June 1978.

Chang, M.S.; Chen, J.T.
Plasma etching of inorganic resists.
Journal of Electronic Materials 8 (5): 727, 1979.

Chang, R.P.H.; Chang, C.C.; Darack, S.
Hydrogen plasma etching of semiconductors and their oxides.
Journal of Vacuum Science and Technology 20 (1): 45–50, 1982.

Chang, R.P.H.; Chang, C.C.; Darack, S.
Hydrogen plasma etching of semiconductors and their oxides.
Journal of Vacuum Science and Technology 20 (3): 490–491, 1982.

Chang, R.P.H.; Darack, S.
Hydrogen plasma etching of GaAs oxide.
Applied Physics Letters 38 (11): 898–900, 1 June 1981.

Chapman, B.N.
Triode systems for plasma etching.
IBM Technical Disclosure Bulletin 21 (12): 5006–5007, May 1979.

Chapman, B.N.
Voltage control in high pressure diode reactors (surface treatment of silicon).
IBM Technical Disclosure Bulletin 22 (12): 5316–5317, May 1980.

Chapman, B.N.; Coburn, J.W.; Hiraoka, H.; Minkiewicz, V.J.; Welsh, L.W.
Plasma etching of a positively biased wafer.
IBM Technical Disclosure Bulletin 22 (3): 1175–1176, Aug. 1979.

Chapman, B.N.; Hansen, T.A.; Minkiewicz, V.J.
Implications of flow rate dependencies in plasma etching.
Journal of Applied Physics 51 (7): 3608–3613, July 1980.

Chapman, B.N.; Heiman, N.; Minkiewicz, V.J.
High rate triode plasma etching of Si.
IBM Technical Disclosure Bulletin 21 (12): 5001, May 1979.

Chapman, B.N.; Minkiewicz, V.J.
Flow rate effects in plasma etching.
Journal of Vacuum Science and Technology 15 (2): 329–332, Mar.–Apr. 1978.

Chapman, B.N.; Nowak, M.
Troublesome aspects of aluminum plasma etching.
Semiconductor International 3 (10): 8 p. between p. 139 and 152, Nov. 1980.

Chen, X.; Pfender, E.
Heat transfer to a single particle exposed to a thermal plasma.
Plasma Chemistry and Plasma Processes 2 (2): 185–212, June 1982.

Chow, T.P.; Steckl, A.J.
Plasma etching characteristics of sputtered $MoSi_2$ films.
Applied Physics Letters 37 (5): 466–468, Sept. 1980.

Chung, S.
Determining a production plasma etch cycle (IC processing).
Solid State Technology 21 (4): 114–116, Apr. 1978.

Clark, H.A.
Plasma etching of SiO_2 polysilicon composite film.
IBM Technical Disclosure Bulletin 20 (4): 1386, Sept. 1977.

Clark, H.A.; Gati, G.S.
Planarization process using a plasma etch.
IBM Technical Disclosure Bulletin 20 (10): 3890–3891, Mar. 1978.

Coburn, J.W.
Enhancing the fragmentation of molecular species in a plasma etching discharge.
IBM Technical Disclosure Bulletin 20 (1): 363, June 1977.

Coburn, J.W.
In situ Auger electron spectroscopy of Si and SiO_2 surfaces plasma etched in CF_4–H_2 glow discharges.
Journal of Applied Physics 50 (8): 5210–5213, Aug. 1979.

Coburn, J.W.
Increasing the selectivity of the plasma etch rate of SiO_2 relative to Si.
IBM Technical Disclosure Bulletin 20 (2): 757, July 1977.

Coburn, J.W.; Chen, M.
Optical emission spectroscopy of reactive plasmas: A method for correlating emission intensities to reactive particle density.
Journal of Applied Physics 51 (6): 3134–3136, June 1980.

Coburn, J.W.; Kay, E.
Some chemical aspects of the fluorocarbon plasma etching of silicon and its compounds.
Solid State Technology 22 (4): 117–124, Apr. 1979.

Coburn, J.W.; Winters, H.F.
Ion and electron assisted gas–surface chemistry – an important effect in plasma etching.
Journal of Applied Physics 50 (5): 3189–3196, May 1979.

Coburn, J.W.; Winters, H.F.
Mechanisms in plasma etching.
Journal of Vacuum Science and Technology 15 (2): 327–328, Mar.–Apr. 1978.

Coburn, J.W.; Winters, H.F.; Chuang, T.J.
Ion surface interactions in plasma etching.
Journal of Applied Physics 48 (8): 3532–3540, Aug. 1977.

Coldren, L.A.; Furuya, K.; Miller, B.I.; Rentschler, J.A.
Combined dry and wet etching techniques to form planar (011) facets in GaInAsP/InP double heterostructures.
Electronics Letters 18 (5): 235–237, Mar. 4, 1982.

Coldren, L.A.; Furuya, K.; Miller, B.I.; Rentschler, J.A.
Etched mirror and groove-coupled GaInAsP/InP laser devices for integrated optics.
IEEE Journal of Quantum Electronics QE-18 (10): 1679–1688, Oct. 1982.

Curran, J.E.
Production of surface patterns by chemical and plasma etching.
Journal of Physics E (Scientific Instruments) 14 (4): 393–407, Apr. 1981.

D'Agostino, R.; Flamm, D.L.
Plasma etching of Si and SiO_2 in SF_6O_2 mixtures.
Journal of Applied Physics 52 (1): 162–167, Jan. 1981.

Danesh, P.; Pantchev, B.G.
Dry etching of crystalline quartz in a planar plasma reactor.
Thin Solid Films 88 (4): 347–352, Feb. 26, 1982.

D'Asaro, L.A.; Butherus, A.D.; DiLorenzo, J.V.; Iglesias, D.E.; Wemple, S.H.
Plasma-etched via connections to GaAs FETS.
Gallium Arsenide and Related Compounds, 8th International Symposium, Vienna, Austria, 22–24 Sept. 1980, p. 267–273.

Devaney, J.R.; Shelbe, J.M.
Plasma etching proms and other problems.
Solid State Technology 17 (12): 46–50, Dec. 1974.

Doken, M.; Miyata, I.
Etching uniformities of silicon in CF_4^+ oxygen plasma.
Journal of the Electrochemical Society 126 (12): 2235–2239, Dec. 1979.

Donnelly, V.M.; Flamm, D.L.
Anisotropic etching in chlorine-containing plasmas.
Solid State Technology 24 (4): 161–166, Apr. 1981.

Donnelly, V.M.; Flamm, D.L.
Studies of chemiluminescence accompanying fluorine atom etching of silicon.
Journal of Applied Physics 51 (10): 5273–5276, Oct. 1980.

Dun, H.; Pan, P.; White, F.R.; Douse, R.W.
Mechanisms of plasma-enhanced silicon nitride deposition using SiH_4–N_2 mixture.
Journal of the Electrochemical Society 128 (7): 1555–1563, July 1981.

Duval, P.
Practical problems with the vacuum pumping in plasma etching and PCVD systems.
Revue de Physique Appliquée 15 (9): 1479–1487, 1980.

Egitto, F.D.; Wang, D.N.K.; Maydan, D.; Benzing, D.
Ion-assisted plasma etching of silicon oxides in a multifacet system.
Solid State Technology 24 (12): 71–75, Dec. 1981.

Eisele, K.M.
Plasma etching of silicon with nitrogen trifluoride.
Journal of the Electrochemical Society 127 (3): PC89, 1980.

Eisele, K.M.
SF_6, a preferable etchant for plasma etching silicon.
Journal of the Electrochemical Society 128 (1): 123–126, Jan. 1981.

Endo, N.; Kurogi, Y.
1 µM MOS process using anisotropic dry etching.
IEEE Journal of Solid-State Circuits SC-15 (4): 411–416, 1980.

Endo, N.; Matsui, S.
Etching characteristics for organosilica.
Japanese Journal of Applied Physics, Pt. 2, 22 (2): L109–111, Feb. 1983.

Enomoto, T.; Denda, M.; Yasuoka, A.; Nakata, H.
Loading effect and temperature dependence of etch rate in CF_4 plasma.
Japanese Journal of Applied Physics 18 (1): 155–163, Jan. 1979.

Flamm, D.L.
Measurements and mechanisms of etchant production during the plasma oxidation of CF_4 and C_2F_6.
Solid State Technology 22 (4): 109–116, Apr. 1979.

Flamm, D.L.
Mechanisms of etchant production and the role of oxidants in CF_3Cl, CF_3Br, and related plasma-etching gases.
Journal of the Electrochemical Society 127 (3): PC87, 1980.

Flamm, D.L.; Ibbotson, D.E.; Mucha, J.A.; Donnelly, V.M.
XeF_2 and F-atom reactions with Si: Their significance for plasma etching.
Solid State Technology 26 (4): 117–121, Apr. 1983.

Flamm, D.L.; Mogab, C.J.; Sklaver, E.R.
Reaction of fluorine atoms with SiO_2.
Journal of Applied Physics 50 (10): 6211–6213, Oct. 1979.

Fok, T.Y.
Plasma etching of aluminum films using CCl_4.
Journal of the Electrochemical Society 127 (3): PC90, 1980.

Fonash, S.J.; Ashok, S.; Singh, R.
Effect of neutral ion beam sputtering and etching on silicon.
Thin Solid Films 90 (3): 231–235, Apr. 23, 1982.

Forster, T.
Dry etch rate control of photoresist by E-beam exposure.
IBM Technical Disclosure Bulletin 23 (8): 3887, Jan. 1981.

Freedman, R.W.; Kaiser, H.D.; Romankiw, L.T.
Selective plasma etching of niobium.
IBM Technical Disclosure Bulletin 20 (4): 1601–1603, Sept. 1977.

Frieser, R.G.; Nogay, J.
Optical spectroscopy applied to the study of plasma etching.
Applied Spectroscopy 34 (1): 31–33, Jan.–Feb. 1980.

Gajda, J.J.; DeLorenzo, D.J.; Wade, J.A.
Failure analysis techniques and failure mechanisms utilizing a plasma etcher.
Semiconductor International 3 (2): 83–90, Feb. 1980.

Garton, A.; Sturgeon, P.Z.; Carlsson, D.J.; Wiles, D.M.
Plasma etching of polypropylene films and fibres.
Journal of Materials Science 13 (10): 2205–2210, Oct. 1978.

Gill, M.D.
A simple technique for monitoring undercutting in plasma etching.
Solid State Electronics 23 (9): 995, Sept. 1980.

Goldstein, I.S.; Kalk, F.
Oxygen plasma etching of thick polymer layers.
Journal of Vacuum Science and Technology 19 (3): 743–747, Sept.–Oct. 1981.

Griffin, S.T.; Verdeyen, J.T.
Plasma processes involved in dry processing.
IEEE Transactions on Electron Devices ED-27 (3): 602–604, Mar. 1980.

Hamamoto, M.
The application of gas plasma for the etching of Si_3N_4 film and its effects on photoresist.
Electrochemical Society Fall Meeting (Extended Abstracts), Dallas, Texas, 5–10 Oct. 1975, p. 335.

Hampy, R.E.
RF plasma enhanced vapor deposition and etching of silicon nitride.
U.S. National Technical Information Service, Report Number SAND-77-2070, 1978.

Harada, T.; Gamo, K.; Namba, S.
Dry etching of Nb and fabrication of Nb variable thickness bridges.
Japanese Journal of Applied Physics 20 (1): 259–264, Jan. 1981.

Hasan, T.F.; Perloff, D.S.
Automated wafer mapping for characterizing plasma etching processes.
Journal of the Electrochemical Society 126 (8): PC373, 1979.

Hayes, J.; Pandhumsoporn, T.
Planar plasma etching of polycrystalline silicon.
Solid State Technology 23 (11): 71–78, Nov. 1980.

Heinecke, R.A.H.
Control of relative etch rates of SiO_2 and Si in plasma etching.
Solid State Electronics 18 (12): 1146–1147, Dec. 1975.

Heinecke, R.A.H.
Plasma etching of films at high rates (IC Al films etching).
Solid State Technology 21 (4): 104–106, Apr. 1978.

Hendricks, C.J.; Shepard, A.K.
Localized variable orifice for plasma etching and deposition system.
IBM Technical Disclosure Bulletin 21 (11): 4469–4471, Apr. 1979.

Herndon, T.O.; Burke, R.L.
Plasma etching. Aluminum. Dry vs. wet.
Circuits Manufacturing 18 (4): 39–40, Apr. 1978.

Heslop, C.J.
Application of reactive plasma processing to integrated circuit manufacture.
International Conference on Ion Plating and Allied Technology, London, England, July 1979, p. 63–70.

Heslop, C.J.
Plasma etching (VLSI processing).
International Electronic Packaging and Production Conference, Proceedings of the Technical Programme, Brighton, England 14-16 Oct. 1980, p. 310–314.

Hess, D.W.
Plasma etching of aluminum.
Solid State Technology 24 (4): 189–194, Apr. 1981.

Hidalgo, M.; Bersin, R.
Use of plasma in the fabrication of rigid and flexible printed circuits.
National Electronic Packaging and Production Conference, Anaheim, California, 15–17 May 1979, p. 251-257.

Hijikta, I.
Plasma etching device capable of handling 64K RAM.
JEE 18 (171): 74–76, Mar. 1981.

Hikosaka, K.; Mimura, T.; Joshin, K.
Selective dry etching of AlGaAs–GaAs heterojunction.
Japanese Journal of Applied Physics 20 (11): L847–850, Nov. 1981.

Hirata, K.; Ozaki, Y.; Oda, M.; Kimizuka, M.
Dry etching technology for 1-μM VLSI fabrication.
IEEE Transactions on Electron Devices ED-28 (11): 1323–1331, 1981.

Hirobe, K.; Kureishi, Y.; Tsuchimoto, T.
Some problems in plasma etching of Al and Al–Si alloy films.
Journal of the Electrochemical Society 128 (12): 2686–2688, Dec. 1981.

Hirobe, Y.; Iwamatsu, S.
Plasma etching on platinum film.
U.S. National Technical Information Service, Report Number SAND-77-6022, 1977.

Hoffman, H.S.; Owen, C.J.
Plasma etching of chrome from cermet resistors.
IBM Technical Disclosure Bulletin 21 (12): 4806, May 1979.

Hofmann, D.; Meier, S.; Wechsung, R.
New spectrometer for investigation and control of plasma processes by plasma ion analysis.
International Vacuum Congress, 8th, Cannes, France, 22–26 Sept. 1980, v. 1, p. 54–57.

Hollahan, J.R.; Wauk, M.T.; Rosler, R.S.
Plasma-enhanced chemical vapor deposition of thin films and some of their etching characteristics.
International Conference on Chemical Vapor Deposition, 6th, Atlanta, Georgia, 9–14 Oct. 1977, p. 224–234.

Horiike, Y.; Okano, H.; Shibagaki, M.
Advanced etching process.
Microcircuit Engineering '82, International Conference on Microlithography, Grenoble, France, 5–8 Oct. 1982.

Horiike, Y.; Shibagaki, M.
A new chemical dry etching.
Japanese Journal of Applied Physics 15 (Suppl. 15-1): 13–18, 1976.

Horiike,; Shibagaki, M.
Dry etching technology using long-lived active species excited by microwave.
Journal of the Electrochemical Society 124 (3): PC120, 1977.

Horiike, Y.; Sugawara, T.; Okano, H.; Shibagaki, M.; Ueda, Y.
Cl_2/Ar plasma etching of a contaminated layer on Si induced by fluorocarbon gas plasma.
Japanese Journal of Applied Physics 20 (4): 803–804, Apr. 1981.

Humphrey, B.L.
Use of oxidized silicon nitride as an etch stop for plasma etching.
IBM Technical Disclosure Bulletin 23 (4): 1360, Sept. 1980.

Hutt, M.
Etching technology.
Sputtering and Plasma Etching Conference, Pebble Beach, California, 10–12 Dec. 1979, 45 p.

Hutt, M.; Class, W.
Optimization and specification of dry etching processes.
Solid State Technology 23 (3): 92–97, Mar. 1980.

Ibbotson, D.E.; Flamm, D.L.; Donnelly, V.M.
Chemical and discharge effects in plasma etching with freon and other electronegative gases.
IEEE Transactions on Electrical Insulation EI-17 (2): 163–167, 1982.

Ibbotson, D.E.; Flamm, D.L.; Donnelly, V.M.; Duncan, B.S.
Studies of plasma etching of III–V compounds – the effects of temperature and discharge frequency.
Journal of Vacuum Science and Technology 20 (3): 489–490, 1982.

Irene, E.A.; Tierney, E.; Blum, J.M.; Aliotta, C.F.; Lamberti, C.; Ginsberg, B.J.
An electron microscope investigation of the effect of phosphorous doping on the plasma etching of polycrystalline silicon.
Journal of the Electrochemical Society 128 (9): 1971–1974, Sept. 1981.

Irving, S.M.; Hayes, J.
Plasma etching. Silicon and dielectric films. Planar vs. barrel etchers.
Circuits Manufacturing 18 (4): 27–28, 30, 33–34, 36, Apr. 1978.

Itakura, H.; Komiya, H.; Toyoda, H.
Plasma etching of SiO_2 relief having tapered wall.
Japanese Journal of Applied Physics 19 (7): 1429–1430, 1980.

Itakura, H.; Komiya, H.; Ukai, K.
Multi-chamber dry etching system.
Solid State Technology 25 (4): 209–214, Apr. 1982.

Jacob, A.
Novel selective RF plasma etching of silicon and silicon-containing compounds encountered in semiconductor manufacture.
Electrochemical Society Fall Meeting (Extended Abstracts), Las Vegas, Nevada, 17–22 Oct. 1976, p. 917–919.

Jacob, A.
Plasma etching uniformity limitations in downstream-of-the-discharge zone processing.
Journal of the Electrochemical Society 126 (8): PC373, 1979.

Jacob, A.
Versatile tecnique of RF plasma etching – 1.
Solid State Technology 19 (9): 70–73, Sept. 1976.

Jacob, A.
Versatile technique of RF plasma etching – 2. Kinetics of etchant formation.
Solid State Technology 20 (6): 31–36, June 1977.

Jacob, A.
Versatile technique of RF plasma etching – 3. Mechanistic considerations for selective etching.
Solid State Technology 21 (4): 95–98, Apr. 1978.

Jankuj, J.
Plasma etching of silicon and its compounds in the freon plasma.
Acta Physica Slovaca 29 (2): 155–199, 1979.

Jawitz, M.; Nellis, E.
Plasma etching rigidflex multilayer boards – a limited study.
National Electronics Packaging and Production Conference, Proceedings of the Technical Program, Anaheim, California, 26–28 Feb. 1980, v. 1, p. 259–281.

Jieping, H.; Xiuying, L.; Zhankun, Y.; Xingcai, X.
A new dry etching technology for metal film.
Chinese Journal of Semiconductors 1 (3): 247–8, Aug. 1980.

Jinno, K.
The etch resistance of electron beam resists in chemical dry etching system using microwave excitation.
Japanese Journal of Applied Physics 17 (7): 1283–1284, July 1978.

Jinno, K.; Kinoshita, H.; Matsumoto, Y.
Etching characteristics of silicate glass films in CF_4 plasma.
Journal of the Electrochemical Society 124 (8): 1258–1262, Aug. 1977.

Jinno, K.; Matsumoto, Y.; Inomata, S.
Characteristics of CF_4 plasma etching.
Denki Kagaku 44 (3): 204–210, 1976.

Jones, N.
Advanced sputtering and plasma etching required for VLSI.
Sputtering and Plasma Etching Conference, Pebble Beach, California, 10–12 Dec. 1979, 30 p.

Jones, W.K.
Plasma etching as applied to failure analysis.
Reliability Physics Symposium, 12th Annual, Las Vegas, Nevada, 2–4 Apr. 1974, p. 43–47.

Kalter, H.; Van de Ven, E.P.G.T.
Plasma etching in IC technolgy.
Philips Technical Review 38 (7–8): 200–210, 1978–1979.

Kalter, H.; Vandijk, V.H.
Feasibility of plasma etching in a barrel reactor.
Journal of the Electrochemical Society 125 (8): PC353, 1978.

Kammerdiner, L.
Aluminum plasma etch considerations for VLSI production.
Solid State Technology 24 (10): 79–85, Oct. 1981.

Kane, S.M.; Tao, L.J.
Etching technique.
IBM Technical Disclosure Bulletin 19 (10): 4017, Mar. 1977.

Kawabe, M.; Kubota, M.; Masuda, K.; Namba, S.
Microfabrication in $LiNbO_3$ by ion-bombardment-enhanced etching.
Journal of Vacuum Science and Technology 15 (3): 1096–1098, May–June 1978.

Kawamoto, Y.; Hashimoto, N.
Plasma etching monitor by electric probe.
Japanese Journal of Applied Physics 18 (Suppl. 1): 277–280, 1979.

Kay, E.; Dilks, A.
Metal-containing fluoropolymer films produced by simultaneous plasma etching and polymerization: The series of perfluoroalkanes C_nF_{2n+2} ($n = 1, 2, 3, 4$).
Thin Solid Films 78 (4): 309–318, Apr. 24, 1981.

Kay, E.; Dilks, A.
Plasma polymerization of fluorocarbons in RF capacitively coupled diode system.
Journal of Vacuum Science and Technology 18 (1): 1–11, Jan.–Feb. 1981.

Kay, E.; Dilks, A.; Seybold, D.
Metal-containing fluoropolymer films produced by simultaneous plasma etching and polymerization: Effects of hydrogen or oxygen.
Journal of Applied Physics 51 (11): 5678–5687, Nov. 1980.

Kim, S.U.
Method of removing nitride overhang ledge by differential etch technique.
IBM Technical Disclosure bulletin 21 (4): 1369–1370, Sept. 1978.

Kitcher, J.R.
Application of plasma etching to multilevel metal structures.
Journal of the Electrochemical Society 127 (8): PC375, 1980.

Kluge, H.C.
Photoresist material for plasma etching of via holes into sputtered silicon dioxide.
IBM Technical Disclosure Bulletin 17 (11): 3270, Apr. 1975.

Koike, A.; Imai, K.; Hosoda, S.; Agatsuma, T.
Anisotropic plasma etching of polysilicon.
Journal of the Electrochemical Society 129 (3): PC105, 1982.

Kominiak, G.J.; Mattox, D.M.
Reactive plasma cleaning of metals.
Thin Solid Films 40 (1–3): 141–148, Jan. 1977.

Komiya, H.; Toyoda, H.; Kato, T.; Inaba, K.
Microfabrication technique by gas plasma etching method.
Japanese Journal of Applied Physics 15 (Suppl. 15-1): 19–24, 1976.

Koste, W.W.; Mathad, G.S.; Patnaik, B.
Via profiling by plasma etching with varying ion energy.
IBM Technical Disclosure Bulletin 22 (7): 2737–2738, Dec. 1979.

Kumar, R.; Ladas, C.; Hudson, G.
Characterization of plasma etching for semiconductor applications.
Solid State Technology 19 (10): 54–59, Oct. 1976.

Kushner, M.J.
A kinetic study of the plasma-etching process. 1. A model for the etching of Si and SiO_2 in C_nF_m/H_2 and C_nF_m/O_2 plasmas.
Journal of Applied Physics 53 (4): 2923–2938, 1982.

Kushner, M.J.
A kinetic study of the plasma-etching process. 2. Probe measurements of electron properties in an RF plasma-etching reactor.
Journal of Applied Physics 53 (4): 2939–2946, 1982.

Lam, D.K.
Role of substrate temperature in plasma processing.
Australian Biochemical Society Proceedings, v. 9, PC285, 1976.

Lam, D.K.; Koch, G.R.
Vacuum system considerations for plasma etching equipment.
Solid State Technology 23 (9): 99–101, Sept. 1980.

Laporte, M.Ph.; Chevalier, M.; Peccoud, L.
Etching of thin films of aluminum by reactive plasma.
International Congress on Cathodic Sputtering and Related Applications, 3rd, Nice, France, 11–14 Sept. 1979, p. 319–341.

Layet, J.M.; Gautherin, G.
Cleaning of semiconductor surfaces with low-energy ions.
International Congress on Cathodic Sputtering and Related Applications, 3rd, Nice, France, 11–14 Sept. 1979, p. 373–381.

LeClaire, R.
Advances in planar plasma etching equipment.
Solid State Technology 22 (4): 139–142, Apr. 1979.

Lee, C.L.; Lu, C.L.
CF_4 plasma etching on $LiNbO_3$.
Applied Physics Letters 35 (10): 756–758, 15 Nov. 1979.

Legat, W.H.; Schilling, H.
Plasma etching of metal films in fabrication of large-scale integrated circuits.
Journal of the Electrochemical Society 122 (8): PC252, 1975.

Lehmann, H.W.; Widmer, R.
Dry etching for pattern transfer.
Journal of Vacuum Science and Technology 17 (5): 1177–1183, Sept.–Oct. 1980.

Lohner, T.; Valyy, G.; Mezey, G.; Kolai, E.
The role of surface cleaning in the ellipsometric studies of ion-implanted silicon.
Journal of Radiation Effects 54 (3/4): 252–252, Feb. 1981.

Ma, W.H.-L.; Ma, T.-P.
RF annealing: A method of removing radiation damage in MIS structures.
International Electron Devices Meeting, Washington, D.C., 5–7 Dec. 1977, p. 151–153.

Maddox, R.L.; Parker, H.L.
Applications of reactive plasma practical microelectronic processing systems.
Solid State Technology 21 (4): 107–113, Apr. 1978.

Mader, H.
Anisotropic plasma etching of polysilicon with CF_4.
Journal of the Electrochemical Society 127 (3): PC88, 1980.

Makino, T.; Nakamura, H.; Asano, M.
Acceleration of plasma etch rate caused by alkaline residues.
Journal of the Electrochemical Society 128 (1): 103–106, Jan. 1981.

Maleham, J.; Armstrong, N.P.
Anisotropic plasma etching of doped polysilicon.
Colloquium on Dry Etching Related to Silicon, London, England, 1–3 June 1982, 1982. Published by IEE, London, England.

Markstein, H.W.
New developments in plasma etching equipment.
Electronic Packaging and Production 18 (5, part 1): 55–58, 60–61, May 1978.

Markstein, H.W.
Plasma etching for desmearing and etchback.
Electronic Packaging and Production 20 (1): 65–66, 68, Jan. 1980.

Markstein, H.W.
Wafer etching: Manual or automated? Wet or dry?
Electronic Packaging and Production 16 (2): 6 p. between p. 24 and 32, Feb. 1976.

Matsuo, S.
Selective etching of Si relative to SiO_2 without undercutting by $CBrF_3$ plasma.
Applied Physics Letters 36 (9): 768–770, 1 May 1980.

Matsuo, S.; Takehara, Y.; Ozawa, A.
Effect of dark space in RF glow discharge on plasma etching characteristics.
Japanese Journal of Applied Physics 17 (11): 2071–2072, Nov. 1978.

Mayer, T.M.
Chemical conversion of C_2F_6 and uniformity of etching SiO_2 in a radial flow plasma reactor.
Journal of Electronic Materials 9 (3): 513–523, May 1980.

Mayer, T.M.; McConville, J.H.
Linewidth control in anisotropic plasma etching of polycrystalline silicon.
International Electron Devices Meeting, 25th, Technical Digest, Washington, D.C., 3–5 Dec. 1979, p. 44–46.

McPherson, R.
Plasma processing of ceramics.
Australian Ceramic Society Journal 17 (1): 2–5, 1981.

Meguro, T.; Kurita, H.; Itoh, T.
Evaluation of the pinhole density in SiO_2 film by plasma etching.
Journal of the Electrochemical Society 128 (6): 1379–1381, June 1981.

Mei, L.; Chen, S.; Dutton, R.W.
A surface kinetics model for plasma etching.
International Electron Devices Meeting, 26th, Technical Digest, Washington, D.C., 8–10 Dec. 1980, p. 831–832.

Miller, S.P.; Stigall, R.E.; Shreve, W.R.
Plasma etched quartz saw resonators.
Ultrasonics Symposium, Los Angeles, California, 22–24 Sept. 1975, p. 474–477.

Mimura, Y.
Fine pattern fabrication using freon gas plasma etching.
Japanese Society of Precision Engineering Journal 42 (6): 485–492, June 1976.

Mimura, Y.
Freon gas plasma dry etching.
Electrical Communication Laboratory Technical Journal 25 (10): 1623–1632, 1976.

Minkiewicz, V.J.; Chapman, B.N.
Triode plasma etching.
Applied Physics Letters 34 (3): 192–193, 1 Feb. 1979.

Minkiewicz, V.J.; Chen, M.; Coburn, J.W.; Chapman, B.N.; Lee K.
Magnetic field control of reactive plasma etching.
Applied Physics Letters 35 (5): 393–394, 1 Sept. 1979.

Mito, H.
Dry etching aids in realization of VLSIs.
JEE 18 (177): 96–98, Sept. 1981.

Mogab, C.J.
The loading effect in plasma etching.
Journal of the Electrochemical Society 124 (8): 1262–1268, Aug. 1977.

Mogab, C.J.; Adams, A.C.; Flamm, D.L.
Plasma etching of Si and SiO_2 – effect of oxygen additions to CF_4 plasmas.
Journal of Applied Physics 49 (7): 3796–3803, 1978.

Mogab, C.J.; Harshbarger, W.R.
Plasma-assisted etching for pattern transfer (IC fabrication).
Journal of Vacuum Science and Technology 16 (2): 408–409, Mar.—Apr. 1979.

Mogab, C.J.; Harshbarger, W.R.
Plasma processes set to etch finer lines with less undercutting.
Electronics 51 (18): 117–121, 31 Aug. 1978.

Mogab, C.J.; Levinstein, H.J.
Anisotropic plasma etching of polysilicon.
Journal of Vacuum Science and Technology 17 (3): 721–730, May–June 1980.

Mogab, C.J.; Shankoff, T.A.
Plasma etching of titanium for application to the patterning of Ti–Pd–Au metallization.
Journal of the Electrochemical Society 124 (11): 1766–1771, Nov. 1977.

Murarka, S.P.; Mogab, C.J.
Contamination of silicon and oxidized silicon wafers during plasma etching.
Journal of Electronic Materials 8 (6): 763–779, Nov. 1979.

Nakamura, M.; Itoga, M.; Ban, Y.
Investigation of aluminum plasma etching by some halogenized gases.
Journal of the Electrochemical Society 127 (3): PC90, 1980.

Nakata, H.; Nishioka, K.; Abe, H.
Plasma etching characteristics of chromium film and its novel etching mode.
Journal of Vacuum Science and Technology 17 (6): 1351–1357, Nov.–Dec. 1980.

Niebauer, D.A.
Plasma etching of rigid/flexible PC boards.
Electronic Packaging and Production 20 (9): 153–154, 156, 158, Sept. 1980.

Nishizawa, J.; Hayasaka, N.
In situ observation in silicon plasma etching.
Journal of the Electrochemical Society 127 (3): PC91, 1980.

Nishizawa, J.; Hayasaka, N.
In situ observation of plasmas for dry etching by IR spectroscopy and probe methods.
Thin Solid Films 92 (1-2): 189–198, June 11, 1982.

Nusta, H.H.; Lajos, R.E.; Kiewit, D.A.
Control plasma etching.
Industrial Research/Development 20 (6): 133, 135, 138, 140, 142, June 1978.

Oda, M.; Hirata, K.
Undercutting phenomena in Al plasma etching.
Japanese Journal of Applied Physics 19 (7): L405–408, July 1980.

O'Hanlon, J.F.
Mechanical pump fluids for plasma deposition and etching systems.
Solid State Technology 24 (10): 86–89, Oct. 1981.

Ohkuma, T.; Mitsui, K.; Inoue, M.
Plasma etching of aluminum and its alloys.
Denki Kagaku 49 (4): 240–244, 1981.

Ojha, S.M.
RF discharge plasma conditions in a plasma processing apparatus.
Vacuum 27 (2): 65–67, Feb. 1977.

Okazaki, S.; Chow, T.P.; Steckl, A.J.
Edge-defined patterning of hyperfine refractory metal silicide MOS structures.
IEEE Transactions on Electron Devices ED-28 (11): 1364–1368, Nov. 1981.

Oldham, W.G.; Neureuther, A.R.; Reynolds, J.L.; Nandgaonkar, S.N.; Chiakang, S.
A general simulator for VLSI lithography and etching processes. II. Application to deposition and etching.
IEEE Transactions on Electron Devices ED-27 (8): 1455–1459, Aug. 1980.

O'Neill, T.G.
Dry etching systems for planar processing.
Semiconductor International 4 (4): 12 p. between p. 67 and 89, Apr. 1981.

Oshima, M.
A study of dry etching-related contaminations on Si and SiO_2.
Surface Science 86 (11): 858–865, July (II) 1979.

Patel, K.V.
Plasma etching of quartz and metals.
IBM Technical Disclosure Bulletin 20 (6): 2200–2201, Nov. 1977.

Porter, R.A.; Harshbarger, W.R.; Clemens, J.T.; Cuthbert, J.D.
Plasma etching of phosphosilicate glass.
Journal of the Electrochemical Society 125 (8): PC353, 1978.

Poulsen, B.
Plasma etching: A new technique for integrated circuit fabrication.
Telesis 5 (3): 73–79, June 1977.

Poulsen, R.G.; Nentwich, H.; Ingrey, S.
Plasma etching of aluminum.
International Electron Devices Meeting (Technical Digest), Washington, D.C., 6–8 Dec. 1976, p. 205–208.

Ranadive, D.K.; Losee, D.L.
Plasma etching of aluminum using CCl_4.
Journal of the Electrochemical Society 127 (3): PC90, 1980.

Rao, P.V.S.; Varker, C.J.
Role of etching in epitaxial growth of silicon.
Electrochemical Society Fall Meeting (Extended Abstracts), Dallas, TX, 5–10 Oct. 1975, p. 436.

Reichelderfer, R.; Vogel, D.; Bersin, R.L.
Plasma etching of aluminum.
Australian Biochemical Society Proceedings, 9: PC286, 1976.

Reinberg, A.R.
Plasma processing with a planar reactor.
Circuits Manufacturing 19 (4): 25–26, Apr. 1979.

Reynolds, J.L.; Neureuther, A.R.; Oldham, W.G.
Simulation of dry etched line edge profiles.
Journal of Vacuum Science and Technology 16 (6): 1772–1775, Nov.–Dec. 1979.

Robb, F.Y.
High resolution polysilicon etching.
Semiconductor International 2 (10): 60–65, Dec. 1979.

Ronsheim, P.; Pfender, E.; Toth, L.
Thermal plasma processing of carbides.
American Ceramic Society Bulletin 60 (8): 851, Aug. 1981.

Rozich, W.R. Jr.; Vrba, E.A.
Determining trace contaminants in an inductively-coupled plasma etching system.
IBM Technical Disclosure Bulletin 20 (3): 1021, Aug. 1977.

Saeki, H.; Watakabe, Y.; Toyoda, H.; Nakata, H.; Kashiwagi, T.
Chromium etching characteristics using a planar type plasma reactor.
Journal of Electronic Materials 11 (6): 1049–1063, Nov. 1982.

Sakai, Y.; Reynolds, J.L.; Neureuther, A.R.
Topography simulation for dry etching process.
Journal of the Electrochemical Society 129 (3): PC101, 1982.

Samarakone, N.
The problems of plasma etching polysilicon as applied to the SOS process.
Colloquium on Dry Etching Related to Silicon, London, England, 1–3 June, 1982.
Published by IEE, London, England.

Sanders, F.H.M.; Dieleman, J.; Peters, H.J.B.; Sanders, J.A.M.
Selective isotropic dry etching of Si_3N_4 over SiO_2.
Journal of the Electrochemical Society 129 (11): 2559–2561, Nov. 1982.

Sarkozy, R.F.; Campbell, P.A.
Plasma etching 1-µM phosphorous-doped polysilicon geometries.
SPIE Proceedings 275: 203–210, 1981.

Scheble, A.M.; Teel, T.P.; Devaney, J.R.
Plasma etching for SEM and EMP examination of microelectronic devices.
Microbeam Analysis Society Conference, 9th Annual (Summaries), Ottawa, Canada, 22–26 July, 1974, p. 31A.

Shen, Y.D.; Welch, B.M.; Zucca, R.
Dry etching multilevel interconnect process for planar GaAs ICs.
Journal of the Electrochemical Society 127 (3): PC90, 1980.

Smolinsky, G.; Chang, R.P.; Mayer, T.M.
Plasma etching of III–V compound semiconductor materials and their oxides.
Journal of Vacuum Science and Technology 18 (1): 12–16, Jan.—Feb. 1981.

Smolinsky, G.; Gottscho, R.A.; Abys, S.M.
Time-dependent etching of GaAs and InP with CCl_4 or HCl plasmas: Electrode materials and oxidant addition effects.
Journal of Applied Physics 54 (6): 3518–3523, June 1983.

Smolinsky, G.; Mayer, T.M.; Truesdale, E.A.
Plasma etching of silicon and silicon dioxide with hydrogen–fluoride mixtures.
Journal of the Electrochemical Society 127 (3): PC88, 1980.

Somekh, S.
Introduction to ion and plasma etching.
Journal of Vacuum Science and Technology 13 (5): 1003–1007, Sept.–Oct. 1976.

Somekh, S.; Casey, H.C. Jr.; Ilegems, M.
Preparation of high-aspect ratio periodic corrugations by plasma and ion etching.
Applied Optics 15 (8): 1905–1906, Aug. 1976.

Suzuki, H.
New plasma etching equipment provides increased precision.
JEE 16 (153): 48–50, Sept. 1979.

Suzuki, K.; Okudaira, S.; Kanomata, I.
The roles of ions and neutral active species in microwave plasma etching.
Journal of the Electrochemical Society 126 (6): 1024–1028, June 1979.

Suzuki, K.; Okudaira, S.; Nishimatsu, S.; Usami, K.; Kanomata, I.
Fundamental characteristics of microwave plasma-etching.
Journal of the Electrochemical Society 126 (8): PC373, 1979.

Suzuki, K.; Okudaira, S.; Sakudo, N.; Kanomata, I.
Microwave plasma etching.
Japanese Journal of Applied Physics 16 (11): 1979–1984, Nov. 1977.

Szekeres, A.: Alexandrova, S.; Kirov, K.
The effect of O_2 plasma on properties of the $Si-SiO_2$ system.
Physica Status Solidi A 62(2): 727–736, 16 Dec. 1980.

Taillet, J.
Ion energy in RF plasma etching.
Comptes Rendus Hebdomadaires des Séances de l'Académie des Sciences, Series B, 287 (16): 325–328, 1978.

Taillet, J.
Plasma physics: Ion energy in RF plasma etching.
Journal of Physics Letters 40 (11): L223–L225, 1 June 1979.

Takahashi, S.; Murai, F.; Asai, S.; Kodera, H.
Reproducible submicron gate fabrication of GaAs FET by plasma etching.
International Electron Devices Meeting (Technical Digest), Washington, D.C., 6–8 Dec. 1976, p. 214–217.

Takahashi, S.; Murai, F.; Kodera, H.
Submicron gate fabrication of GaAs MESFET by plasma etching.
IEEE Transactions on Electron Devices ED-25 (10): 1213–1218, Oct. 1978.

Takahashi, S.; Murai, F.; Kurono, H.; Hirao, M.; Kodera, H.
Half-micron gate GaAs FET fabricated by chemical dry etching.
Japanese Journal of Applied Physics 16 (Suppl. 16-1): 115–118, 1977.

Taylor, W.E.; Bunyan, S.M.; Olson, C.E.
Effects of plasma etching solar-cell front surfaces.
Journal of the Electrochemical Society 127 (3): PC108, 1980.

Teii, S.; Matsumura, S.; Ichikawa, Y.; Hobson, R.M.; Fukuda, H.; Chang, J.S.
The diffusion plasma columns in a coaxial cylindrical glow discharge tube and its application to plasma discharge chambers.
International Conference on Gas Discharges and Their Applications, 7th, London, England, 31 Aug.–3 Sept. 1982, p. 363–366.

Theeten, J.B.
Real-time and spectroscopic ellipsometry of film growth: Application to multilayer systems in plasma and CVD processing of semiconductors.
Surface Science 96 (1/3): 275–293, June 1980.

Tiller, H.J.; Apfel, K.; Voigt, R.
Problems of surface morphology and layer deposition during plasma etching processes.
Crystal Research and Technology 16 (11): PK133, 1981.

Tokunaga, K.; Hess, D.W.
Aluminum etching in carbon tetrachloride plasmas.
Journal of the Electrochemical Society 127 (4): 928–932, Apr. 1980.

Tokunaga, K.; Hess, D.W.
Plasma-etching of aluminum films in carbon tetrachloride.
Journal of the Electrochemical Society 126 (8): PC373, 1979.

Tokunaga, K.; Redeker, F.C.; Danner, D.A.; Hess, D.W.
Comparison of aluminum etch rates in carbon tetrachloride and boron trichloride plasmas.
Journal of the Electrochemical Society 128 (4): 851–855, Apr. 1981.

Toth, L.E.; Ronsheim, P.; Pfender, E.
Thermal plasma processing of carbides.
Journal of Metals 33 (9): PA17, 1981.

Toyoda, H.; Tobinaga, M.; Komiya, H.
Frequency effect on material selectivity in gas plasma etching in planar type reactor.
Japanese Journal of Applied Physics 20 (3): 681–682, Mar. 1981.

Tsumita, N.; Melngailis, J.; Hawryluk, A.M.; Smith, H.I.
Si and shadowing techniques.
Journal of Vacuum Science and Technology 19 (4): 1211–1213, Nov.–Dec. 1981.

Valyi, G.; Schiller, V.; Gyimesi, J.; Gyulai, J.
Analysis of chemical processes of plasma etching.
Thin Solid Films 76 (3): 215–219, 13 Feb. 1981.

Vandeven, E.P.G.T.; Zijlstra, P.A.
Critical comparison of SiF_4–O_2 and CF_4–O_2 as plasma etching gases.
Journal of the Electrochemical Society 127 (3): PC87, 1980.

Van Roosmalen, A.J.
Plasma parameter estimation from RF impedance measurements in a dry etching system.
Applied Physics Letters 42 (5): 416–418, Mar. 1, 1983.

Vasile, M.J.
Etching of SiO_2 and Si in a He–F_2 plasma.
Journal of Applied Physics 51 (5): 2510–2515, May 1980.

Villamizar, C.A.; Feijod, L.; Miller, A.; Vasquez, P.
Chemical etching versus plasma etching in electroplating polypropylene resin surfaces.
American Chemical Society Abstracts of Papers 182: 187, Aug. 1981.

Villamizar, C.A.; Rojas, J.; Frias, P.
Chemical etching versus plasma etching in electroplating ABS resin surfaces.
Metal Finishing 79 (3): 27–33, Mar. 1981.

Viswanathan, N.S.
Simulation of plasma-etched lithographic structures.
Journal of Vacuum Science and Technology 16 (2): 388–390, Mar.–Apr. 1979.

Voschenkov, A.M.; Bartelt, J.L.
Shielded plasma etching of polysilicon-MOS structures: A C–V evaluation.
Electrochemical Society Fall Meeting (Extended Abstracts), Dallas, Texas, 5–10 Oct. 1975, p. 333–334.

Vossen, J.L.
Glow discharge phenomena in plasma etching and plasma deposition.
Journal of the Electrochemical Society 126 (2): 319–324, Feb. 1979.

Weiss, A.D.
Plasma etching of oxides and nitrides.
Semiconductor International 6 (2): 56–62, Feb. 1983.

White, F.R.; Koburger, C.W.; Geipel, H.J.; Harmon, D.L.
Plasma etching of composite silicide gate electrodes.
Journal of the Electrochemical Society 127 (8): PC375, 1980.

Winters, H.F.
The role of chemisorption in plasma etching.
Journal of Applied Physics 49 (10): 5165–5170, Oct. 1978.

Winters, H.F.; Coburn, J.W.
The etching of silicon with XeF_2 vapor.
Applied Physics Letters 34 (1): 70–73, 1 Jan. 1979.

Winters, H.F.; Coburn, J.W.; Kay, E.
Plasma etching – a 'pseudo-black-box' approach.
Journal of Applied Physics 48 (12): 4973–4983, Dec. 1977.

Wydeven, T.; Johnson, C.C.; Golub, M.A.; Hsu, M.S.; Lerner, N.R.
Plasma etching of poly (N,N'–(P,P'–oxydiphenylene) pyromellitimide film and photo thermal degradation of etched and unetched film.
ACS Symposium Series, v. 1979 (108): 299–314, 1979.

Yamazaki, T.; Suzuki, Y.; Nakata, H.
Gas plasma etching of ion-implanted chromium films.
Journal of Vacuum Science and Technology 17 (6): 1348–1350, Nov.–Dec. 1980.

Yamazaki, T.; Suzuki, Y.; Uno, J.; Nakata, H.
Reversal etching of chromium film in gas plasma.
Journal of the Electrochemical Society 126 (10): 1794–1798, Oct. 1979.

Yamazaki, T.; Watakabe, Y.; Suzuki, Y.; Nakata, H.
A dry etching technique using electron beam resist–PBS.
Journal of the Electrochemical Society 127 (8): 1859–1861, Aug. 1980.

Young, M.Y.T.; Pancholy, R.K.; Hagen, G.
Planar plasma etching applications to CMOS–SOS devices.
Journal of the Electrochemical Society 127 (8): PC392, 1980.

Zafiropoulo, A.
Dry vs. wet plasma etching/stripping comes of age.
Circuits Manufacturing 16 (4): 42, 44, 46, Apr. 1976.

Zelley, A.
Dependence of plasma etch rate and uniformity on resist type and processing.
SPIE Proceedings, v. 174: Developments in Semiconductor Microlithography 4, San Jose, California, 23–24 Apr. 1979, p. 173–117.

4.3. Reactive ion etching

Anonymous.
Production RIE.
Solid State Technology 26 (4): 71–72, Apr. 1983.

Anonymous.
Reactive ion etcher (for VLSI).
Solid State Technology 26 (3): 33–34, Mar. 1983.

Anantha, N.G.; Bhatia, H.S.; Bhatia, S.S.
JFET structure.
IBM Technical Disclosure Bulletin 25 (3A): 984–985, Aug. 1982.

Aritome, H.; Yamato, T.; Matsui, S.; Namba, S.
A blazed Si grating for soft X-rays fabricated by two-stage reactive ion beam etching.
Japanese Journal of Applied Physics Part 2, 22 (4): L219–220, Apr. 1983.

Asakawa, M.; Matsuo, S.; Oshima, M.
Plasma reactive sputter etching system.
Electrical Communications Laboratory Technical Journal 27 (9): 2117–2130, 1978.

Ashok, S.; Chow, T.P.; Baliga, B.J.
Modification of Schottky barriers in silicon by reactive ion etching with NF_3.
Applied Physics Letters 42 (8): 687–689, Apr. 15, 1983.

Barson, F.
ROI isolation process to minimize bird's beak (recessed oxide isolation).
IBM Technical Disclosure Bulletin 25 (8): 4429–4430, Jan. 1983.

Bauer, H.
Eliminating inhomogeneities in sputter or reactive ion etching.
IBM Technical Disclosure Bulletin 23 (5): 1984, Oct. 1980.

Beinvogel, W.; Hasler, B.
Reactive ion etching of polysilicon and tantalum silicide.
Solid State Technology 26 (4): 125–130, Apr. 1983.

Beinvogel, W.; Mader, H.
Reactive dry etching for fabrication of very large scale integrated circuits.
Siemens Forschungs und Entwicklungsberichte/Research and Development Reports 11 (4): 180–189, 1982.

Bennett, R.S.; Dimeo, M.A.; Ephrath, L.M.
Perforated counter-electrode for flexible diode RIE system.
IBM Technical Disclosure Bulletin 23 (6): 2579–2580, Nov. 1980.

Bensaoula, A.; Wolfe, J.C.; Oro, J.A.; Ignatiev, A.
Deposition and reactive ion etching of molybdenum.
Applied Physics Letters 42 (1): 122–123, Jan. 1, 1983.

Beyer, K.D.; Gdula, R.A.
Reduction of oxidation-induced defects using reactive ion etching.
IBM Technical Disclosure Bulletin 22 (3): 679–680, July 1979.

Bhatia, H.S.; Gardiner, J.R.; Pliskin, W.A.; Revitz, M.
Single-step process for forming isolation trenches of different depths in a silicon substrate (reactive ion etching).
IBM Technical Disclosure Bulletin 25 (4): 1890–1891, Sept. 1982.

Bollinger, L.D.
Ion beam etching with reactive gases.
Solid State Technology 26 (1): 99–108, Jan. 1983.

Bondur, J.A.
Dry process technology (reactive ion etching).
Journal of Vacuum Science and Technology 13 (5): 1023–1029, Sept.–Oct. 1976.

Bondur, J.A.; Schwartz, S.M.
Selective reactive ion etching of silicon compounds.
IBM Technical Disclosure Bulletin 21 (10): 4015, Mar. 1979.

Borghesani, A.F.; Valente, M.
The technique of dry etching with reactive gas plasma in the manufacture of semiconductor circuits.
Fisica e Tecnologia 5 (1): 3–15, Jan.—Mar. 1982.

Boyar, S.; Chang, K.; Chiu, G.T.; Kitcher, J.R.
Quartz trench RIE etch stop.
IBM Technical Disclosure Bulletin 24 (10): 5133–5134, Mar. 1982.

Boyd, G.D.; Coldren, L.A.; Storz, F.G.
Directional reactive ion etching at oblique angles.
Applied Physics Letters 36 (7): 583–585, 1980.

Bresnock, F.J.
Implementation of adaptive process control to a dry etching process.
Journal of Vacuum Science and Technology 20 (4): 1027–1030, Apr. 1982.

Cabral, S.M.; Elta, M.E.; Chu, A.; Mahoney, L.J.
Reactive ion etching of gold using SF_6.
Journal of the Electrochemical Society 129 (3): PC105, 1982.

Castellano, R.N.
Reactive sputter etching of thin films for pattern delineation.
IEEE Transactions on Components, Hybrids, and Manufacturing Technology CHMT-1 (4): 397–399, 1978.

Chambers, A.A.
The application of reactive ion etching to the definition of patterns in Al–Si–Cu alloy conductor layers and thick silicon oxide films.
Solid State Technology 26 (1): 83–86, Jan. 1983.

Chambers, A.A.
Reactive ion etching of Al–Si–Cu alloy films.
Solid State Technology 25 (8): 93–97, Aug. 1982.

Chinn, J.D.; Adesida, I.; Wolf, E.D.; Tiberio, R.C.
Reactive ion etching for submicron structures.
Journal of Vacuum Science and Technology 19 (4): 1418–1422, Nov.–Dec. 1981.

Chu, S.F.; Gilmartin, P.A.; Mauer, J.L.; Turene, F.E.
CF_4/H_2 induced diffusion barrier and method of its elimination.
IBM Technical Disclosure Bulletin 24 (4): 2142, Sept. 1981.

Coburn, J.W.; Winters, H.F.
Etching in reactive plasmas.
Journal of Vacuum Science and Technology 16 (6): 1613–1614, Nov.–Dec. 1979.

Cohen, C.
Reactive ion etching goes commercial, promising to boost LSI and VLSI yields.
Electronics 53 (24): 76, 1980.

Coldren, L.A.; Iga, K.; Miller, B.I.; Rentschler, J.A.
GaInAsP/InP stripe-geometry laser with reactive ion etched facet.
Applied Physics Letters 37 (8): 681–683, 15 Oct. 1980.

Coldren, L.A.; Iga, K.; Miller, B.I.; Rentschler, J.A.
Integrated GaInAsP/InP lasers formed by reactive ion etching.
IEEE Transactions on Electron Devices ED-27 (11): 2192, 1980.

Coldren, L.A.; Miller, B.I.; Iga, K.; Rentschler, J.A.
Monolithic two-section GaInAsP/InP active optical resonator devices formed by reactive ion etching.
Applied Physics Letters 38 (5): 315–317, 1981.

Coldren, L.A.; Rentschler, J.A.
Directional reactive ion etching of InP with Cl_2 containing gases.
Journal of Vacuum Science and Technology 19 (2): 225–230, 1981.

Craighead, H.G.; Howard, R.E.; Tennant, D.M.
Textured thin-film silicon solar selective absorbers using reactive ion etching.
Applied Physics Letters 37 (7): 653–655, Oct. 1980.

Crimi, C.F.; Eames, W.S.; Friedman, J.D.; Montillo, F.J.
Increased dielectric strength of thermal oxide films by reactive ion etching.
IBM Technical Disclosure Bulletin 22 (11): 4891, Apr. 1980.

Cuomo, J.J.; Leary, P.A.
Reactive ion etching of copper.
IBM Technical Disclosure Bulletin 25 (12): 6394, May 1983.

Danesh, P.; Pantchev, B.G.
Reactive ion etching of crystalline quartz.
Thin Solid Films 82 (1): PL117, 1981.

Das, G.; Mader, S.R.
Elimination of reactive ion etching trench-induced defects.
IBM Technical Disclosure Bulletin 23 (12): 5344, May 1981.

Das, G.; Montillo, F.J.
Elimination of RIE-induced metallic contamination.
IBM Technical Disclosure Bulletin 23 (10): 4490, Mar. 1981.

Decker, S.K.; Hitchner, J.E.; Williams, J.L.
Reactive ion etching of titanium–permalloy films.
IBM Technical Disclosure Bulletin 22 (12): 5433, May 1980.

De Prost, C.; Fried. T.
Reactive ion etching of polysilicon: Evaluation of single wafer processing machine.
Microcircuit Engineering '82. International Conference on Microlithography, Grenoble, France, 5–8 Oct. 1982.

De Werdt, R.; Meeuwissen, W.A.M.
Structuring of bubble overlays by reactive sputter etching in an Ar/H_2O atmosphere.
Journal of Vacuum Science and Technology 16 (6): 2093–2095, Nov.–Dec. 1979.

Dimaria, D.J.; Ephrath, L.M.; Young, D.R.
Radiation damage in silicon dioxide films exposed to reactive ion etching.
Journal of Applied Physics 50 (6): 4015–4021, June 1979.

Duffy, M.I.; Corboy, J.F.; Soltis, R.A.
Reactive sputter etching of dielectrics.
RCA Review 44 (1): 157–168, Mar. 1983.

Eames, W.S.; Friedman, J.D.
Method for polysilicon profile control using CVD oxide and reactive ion etching.
IBM Technical Disclosure Bulletin 22 (7): 2739, Dec. 1979.

Ephrath, L.M.
Effect of cathode materials on reactive ion etching of Si and SiO_2 in a CF_4 plasma.
Journal of Electronic Materials 7 (3): 415–428, May 1978.

Ephrath, L.M.
Effect of cathode materials on reactive ion etching of Si and SiO_2 in CF_4.
Journal of Electronic Materials 5 (4): 449, 1976.

Ephrath, L.M.
Reactive ion etching for VLSI.
IEEE Transactions on Electron Devices ED-28 (11): 1315–1319, Nov. 1981.

Ephrath, L.M.
Reactor design for selective RIE of SiO_2.
IBM Technical Disclosure Bulletin 25 (10): 5045–5046, Mar. 1983.

Ephrath, L.M.
Selective etching of silicon dioxide using reactive ion etching with CF_4–H_2.
Journal of the Electrochemical Society 126 (8): 1419–1421, 1979.

Ephrath, L.M.
Two-step dry process for delineating micron and submicron dimension polysilicon gates.
IBM Technical Disclosure Bulletin 21 (10): 4236, Mar. 1979.

Ephrath, L.M.; Bennett, R.S.
RIE contamination of blanket etched silicon surfaces.
Journal of the Electrochemical Society 127 (8): PC372, 1980.

Ephrath, L.M.; Bennett, R.S.
RIE contamination of etched silicon surfaces.
Journal of the Electrochemical Society 129 (8): 1822–1826, Aug. 1982.

Ephrath, L.M.; Dimaria, D.J.; Pesavento, F.L.
Parameter dependence of RIE induced radiation damage in silicon dioxide.
Journal of the Electrochemical Society 128 (11): 2415–2419, Nov. 1981.

Ephrath, L.M.; Dimaria, D.J.; Pesavento, F.L.
RIE-induced radiation damage.
Journal of the Electrochemical Society 126 (8): PC338, 1979.

Ephrath, L.M.; Hunter, W.R.; Luhn, H.E.
Contact (via) hole processing using $CF_4 + H_2$ reactive ion etching and liftoff.
IBM Technical Disclosure Bulletin 22 (11): 5167, Apr. 1980.

Ephrath, L.M.; Petrillo, E.J.
Parameter and reactor dependence of selective oxide RIE in $CF_4 + H_2$.
Journal of the Electrochemical Society 129 (10): 2282–2287, Oct. 1982.

Forney, G.B.; Planck, R.E.; Rozich, W.R.
Reduction of emitter-based defects with reactive ion etching.
IBM Technical Disclosure Bulletin 25 (11B): 6122, Apr. 1983.

Foxe, T.T.; Hunt, B.D.; Rogers, C.; Kleinsasser, E.W.; Buhrman, R.A.
Reactive ion etching of niobium.
Journal of Vacuum Science and Technology 19 (4): 1394–1397, 1981.

Fraser, D.B.; Kinsbron, E.; Vratny, F.; Johnston, R.L.
NMOS silicide polysilicon gates by lift-off reactive sputter etching.
Journal of Vacuum Science and Technology 20 (3): 491–492, Mar. 1982.

Garbarino, P.L.; Johnson, C. Jr.; Wilbarg, R.R.
Application of reactive ion etch in fabrication of high performance charge-coupled shift registers.
IBM Technical Disclosure Bulletin 20 (1): 131–132, June 1977.

Gartner, H.M.; Hitchner, J.T.; Hoeg, A.J.; Sarkary, H.G.
Achieving uniform etch rates in reactive ion plasma etching processes.
IBM Technical Disclosure Bulletin 20 (7): 2703, Dec. 1977.

Gartner, H.M.; Hitchner, J.T.; Hoeg, A.J.; Sarkary, H.G.
Isotropic and anisotropic etching in a diode system.
IBM Technical Disclosure Bulletin 20 (5): 1744–1745, Oct. 1977.

Gartner, H.M.; Hoeg, A.J.; Jani, R.K.; Sarkary, H.G.; Tewari, V.
Selective etch rate control technique in reactive ion etching.
IBM Technical Disclosure Bulletin 21 (3): 1032–1033, Aug. 1978.

Gartner, H.M.; Hoeg, A.J.; Sarkary, H.G.; Schmidt, H.W. Jr.
In situ thermal control/monitor system for reactive ion etch process.
IBM Technical Disclosure Bulletin 20 (3): 994, Aug. 1977.

Gdula, R.A.
SF_6 RIE of polysilicon.
Journal of the Electromechanical Society 126 (8): PC373, 1979.

Gdula, R.A.; Hollis, J.C.; Pliskin, W.A.
Method of controlling RIE (reactive ion etching) mesa edge profiles to eliminate mouseholing.
IBM Technical Disclosure Bulletin 21 (6): 2327–2328, Nov. 1978.

Gipstein, E.; Gritter, R.J.; Need, O.U. III; Yoon, D.Y.
Radiation sensitive, high temperature, RIE resistant polymeric resist.
IBM Technical Disclosure Bulletin 20 (3): 1205, Aug. 1977.

Grabbe, P.; Hpward, R.E.; Hu, E.L.; Tennant, D.M.
Metal-on-polymer masks for reactive ion etching.
Journal of the Electrochemical Society 129 (3): PC112, 1982.

Greschner, J.; Mohr, T.; Trumpp, H.J.; Vettinger, R.
Lithographic method for defining edge angles.
IBM Technical Disclosure Bulletin 25 (11B): 6185–6186, Apr. 1983.

Harder, A.R.; Spencer, O.S.
Gas mixture control permits nonselective reactive ion etch.
IBM Technical Disclosure Bulletin 21 (4): 1518–1519, Sept. 1978.

Hazuki, Y.; Moriya, T.; Kashiwagi, M.
A new application of RIE to planarization and edge rounding of SiO_2 hole in the Al multi-level interconnection.
Symposium on VLSI Technology, Digest of Papers, Oiso, Japan, 1–3 Sept. 1982, p. 18–19.

Heiman, N.; Minkiewicz, V.; Chapman, B.
High rate reactive ion etching of Al_2O_3 and Si.
Journal of Vacuum Science and Technology 17 (3): 731–734, May–June 1980.

Heimeier, H.H.; Joy, R.C.; Magdo, I.E.; Phillips, A.R. Jr.
Increased bipolar F_t with one-step emitter etch.
IBM Technical Disclosure Bulletin 25 (3A): 980, Aug. 1982.

Hendricks, C.J.
Cathode plate for backside oxide strip in an RIE reactor.
IBM Technical Disclosure Bulletin 23 (12): 5350–5351, May 1981.

Heslop, C.J.
Reactive plasma processing in IC manufacture.
Electronic Production 9 (2): 44–45, 47, 49, Feb. 1980.

Hikita, M.; Nakamura, K.; Kubo, S.; Igarashi, M.; Kakuchi, M.; Kogure, O.
Fabrication of I–V characteristics of high-$T_c Nb_3Ge$ microbridges.
Japanese Journal of Applied Physics, Part 2, 21 (1): 10–12, Jan. 1982.

Hiraoka, H.; Welsh, L.W.
Polymer degradation in reactive ion etching and its possible application to all dry processes.
Radiation Physics and Chemistry 18 (5-6): 907–911, 1981.

Hirobe, K.; Kureishi, Y.; Tsuchimoto, T.
Effects of surface oxide-films on aluminum reactive sputter etching.
Journal of the Electrochemical Society 126 (8): PC373, 1979.

Hitchner, J.E.; Hoeg, A.J.; Joseph, R.R.; Kitcher, J.R.
Chromium as an RIE etch barrier.
IBM Technical Disclosure Bulletin 22 (10): 4516–4517, Mar. 1980.

Hitchner, J.E.; O'Rourke, G.D.
Polyimide layers having tapered via holes.
IBM Technical Disclosure Bulletin 20 (4): 1384, Sept. 1977.

Horiike, Y.; Okano, H.; Yamazaki, T.; Horie, H.
High-rate reactive ion etching of SiO_2 using a magnetron discharge.
Japanese Journal of Applied Physics 20 (11): PL817–820, Nov. 1981.

Horiike, Y.; Yamazaki, T.; Shibagaki, M.; Kurisaki, T.
Aluminum reactive ion etching employing $CCl_4 + Cl_2$ mixture.
Japanese Journal of Applied Physics, Part 1, 21 (10): 1412–1420, Oct. 1982.

Horwitz, C.M.
Reactive sputter etching of silicon with very low mask-material etch rates.
IEEE Transactions on Electron Devices ED-28 (11): 1320–1323, Nov. 1981.

Horwitz, C.M.; Melngailis, J.
Reactive sputter etching of Si, SiO, Cr, Al, and other materials with gas mixtures based on CF_4 and Cl_2.
Journal of Vacuum Science and Technology 19 (4): 1408–1411, Nov.–Dec. 1981.

Hu, E.L.; Howard, R.E.
Reactive ion etching of GaAs and InP using $CCl_2F_2/Ar/O_2$.
Applied Physics Letters 37 (11): 1022–1024, 1 Dec. 1980.

Jackel, L.D.; Howard, R.E.; Hu, E.L.; Lyman, S.P.
Reactive ion etching of $LiNbO_3$.
Applied Physics Letters 38 (11): 907–909, 1981.

Jackel, L.D.; Howard, R.E.; Hu, E.L.; Tennant, D.M.; Grabbe, P.
50-nm silicon structures fabricated with trilevel electron beam resist and reactive ion etching.
Applied Physics Letters 39 (3): 268–270, 1 Aug. 1981.

Jech, C.
Radioactive recoil implant as a tracer in reactive ion etching studies.
Radiochemical and Radioanalytical Letters 47 (1-2): 83–88, 1981.

Johnson, C.; Patnaik, B.
Polyimide mask for reactive ion etching of metal lines.
IBM Technical Disclosure Bulletin 22 (4): 1446–1448, Sept. 1979.

Johnson, C. Jr.; Wilbarg, R.R.
Etch stop for reactive ion etching of polysilicon.
IBM Technical Disclosure Bulletin 21 (2): 599–600, July 1978.

Kaiser, H.D.; Metzger, W.C.; Ting, C.H.
CCl_4 reactive ion etching of niobium films.
IBM Technical Disclosure Bulletin 23 (4): 1694, Sept. 1980.

Kalk, F.; Glocker, D.
Thick zone plate fabrication using reactive sputter etching.
Journal of Vacuum Science and Technology 19 (2): 170–172, July–Aug. 1981.

Kinsbron, E.; Fraser, D.B.; Vratny, F.
High conductivity silicide on polycrystalline silicon prepared by lift-off reactive sputter etching.
Thin Solid Films 90 (3): 243, Apr. 23, 1982.

Kitcher, J.R.
Reactive ion etch process for metal wiring using a buried mask.
IBM Technical Disclosure Bulletin 23 (4): 1394, Sept. 1980.

Klinger, R.E.; Greene, J.E.
Glow-discharge optical spectroscopy studies of the reactive ion etching of Si.
Journal of the Electrochemical Society 127 (3): PC90, 1980.

Klinger, R.E.; Greene, J.E.
Reactive ion etching of GaAs in CCl_2F_2.
Applied Physics Letters 38 (8): 620–622, 15 Apr. 1981.

Klinger, R.E.; Greene, J.E.
Reactive ion etching of GaAs in $CCl_{4-x}F_x$ ($x = 0, 2, 4$) and mixed $CCl_{4-x}F_x$Ar discharges.
Journal of Applied Physics 54 (3): 1595–1604, Mar. 1983.

Knop, K.; Lehmann, H.W.; Widmer, R.
Diffractive optical filters micro-fabricated by reactive sputter etching.
Journal of the Electrochemical Society 125 (3): PC155, 1978.

Knop, K.; Lehmann, H.W.; Widmer, R.
Microfabrication and evaluation of diffractive optical filters prepared by reactive sputter etching.
Journal of Applied Physics 50 86): 3841—3849, June 1979.

Kolodner, P.; Katzir, A.; Hartsough, N.
Noncontact surface temperature measurement during reactive ion etching using fluorescent polymer films.
Applied Physics Letters 42 (8): 749–751, Apr. 15, 1983.

Krongelb, S.
Patterning of cathode element arrays by reactive ion etching.
IBM Technical Disclosure Bulletin 25 (11B): 5918–5920, Apr. 1983.

Krongelb, S.
Process for patterning SiC by reactive ion etching.
IBM Technical Disclosure Bulletin 23 (2): 828–829, July 1980.

Kuwano, H.; Miyake, S.; Kasai, T.
Dry cleaning of Si surface contamination by reactive sputter etching.
Japanese Journal of Applied Physics, Part 1, 21 (3): 529–533, Mar. 1982.

Lechaton, J.S.; MacPherson, C.C.
Method for forming vertical walled trenches in silicon substrates using reactive sputter etching.
IBM Technical Disclosure Bulletin 25 (8): 4408–4409, Jan. 1983.

Lechaton, J.S.; Srinivasan, G.R.
High–low RIE process.
IBM Technical Disclosure Bulletin 25 (11B): 6118, Apr. 1983.

Lechaton, J.S.; Srinivasan, G.R.; Gaur, S.P.
Precision reactive sputter etching and its applications.
Japanese Journal of Applied Physics Suppl., 141–144, 1982.

Lee, M.H.; Schwartz, G.C.
Reactive ion etching process.
IBM Technical Disclosure Bulletin 22 (8A): 3347, Jan. 1980.

Lee, W.Y.; Chen, M.; Eldridge, J.M.; Schwartz, G.C.
Reactive ion etching induced corrosion of Al and Al–Cu films.
Journal of Vacuum Science and Technology 18 (2): 359–360, 1981.

Lee, W.Y.; Eldridge, J.M.; Schwartz, G.C.
Reactive ion etching induced corrosion of Al and Al–Cu films.
Journal of Applied Physics 52 (4): 2994–2999, Apr. 1981.

Lehmann, H.W.; Widmer, R.
Fabrication of deep square wave structures with micron dimensions by reactive sputter etching.
Applied Physics Letters 32 (3): 163–165, 1 Feb. 1978.

Lehmann, H.W.; Widmer, R.
Profile control by reactive sputter etching.
Journal of Vacuum Science and Technology 15 (2): 319–326, Mar.–Apr. 1978.

Lever, R.F.; Kaplan, L.H.
Resist modification for radiative cooling during reactive ion etching.
IBM Technical Disclosure Bulletin 22 (11): 5154, Apr. 1980.

Light, R.W.; See, F.C.
Reactive ion etching of silicon dioxide.
Journal of the Electrochemical Society 129 (5): 1152–1154, May 1982.

Lincoln, G.A.; Geis, M.W.; Mahoney, L.J.; Chu, A.; Vojak, B.A.; Nichols, K.B.; Piacentini, W.J.; Efremow, N.; Lindley, W.T.
Ion beam assisted etching for GaAs device applications.
Journal of Vacuum Science and Technology 20 (3): 786–789, Mar. 1982.

Logan, J.S.; Mauer, J.L.; Schwartz, G.C.; Zielinski, L.B.
Process for forming tapered vias in SiO_2 by reactive ion etching.
IBM Technical Disclosure Bulletin 22 (1): 130–132, June 1979.

Maa, J.-S.; O'Neill, J.J.
Reactive ion etching of Al and Al–Si films with CCl_4, N_2, and BCl_3 mixtures.
Journal of Vacuum Science and Technology A 1(2, Part 1): 636–637, Apr.–June 1983.

Manzi, M.; Patnaik, B.
Reactive ion etching of polymeric and dielectric materials.
IBM Technical Disclosure Bulletin 25 (11A): 5744–5745, Apr. 1983.

Marinaccio, F.A.; Rozich, W.R. Jr.; Waite, T.W.
Removal of RIE-induced Si_3N_4 ribbons.
IBM Technical Disclosure Bulletin 24 (11A): 5547–5548, Apr. 1982.

Matsui, S.; Yamato, T.; Aritome, H.; Namba, S.
Fabrication of SiO_2 blazed holographic gratings by reactive ion etching.
Japanese Journal of Applied Physics 19 (3): L126–128, Mar. 1980.

Matsumoto, H.; Sugano, T.
Characterization of reactive ion etched silicon surface by deep level transient spectroscopy.
Journal of the Electrochemical Society 129 (12): 2823–2828, Dec. 1982.

Matsuo, S.
Etching characteristics of various materials by plasma reactive sputter etching.
Japanese Journal of Applied Physics 17 (1): 235–236, Jan. 1978.

Matsuo, S.
Selective etching of SiO_2 relative to Si by plasma reactive sputter etching.
Journal of Vacuum Science and Technology 17 (2): 587–594, 1980.

Matsuo, S.; Takehara, Y.
Preferential SiO_2 etching on Si substrate by plasma reactive sputter etching.
Japanese Journal of Applied Physics 16 (1): 175–176, Jan. 1977.

Mattausch, H.J.; Hasler, B.; Beinvogel, W.
Reactive ion etching of Ta-silicide/polysilicon double layers for the fabrication of integrated circuits.
Journal of Vacuum Science and Technology B 1 (1): 15–22, Jan.–Mar. 1983.

Mauer, J.L.; Logan, J.S.
Effects of pressure in reactive ion etching.
Journal of the Electrochemical Society 126 (8): PC373, 1979.

Mauer, J.L.; Logan, J.S.
Reactant supply in reactive ion etching.
Journal of Vacuum Science and Technology 16 (2): 404–406, 1979.

Mauer, J.L.; Logan, J.S.; Zielinski, L.B.; Schwartz, G.C.
Mechanism of silicon etching by a CF_4 plasma.
Journal of Vacuum Science and Technology 15 (5): 1734–1738, Sept.–Oct. 1978.

Mauer, J.L.; Schwartz, G.C.
Mass balance of the reactive ion etching products of silicon on SiO_2.
Journal of the Electrochemical Society 127 (3): PC88, 1980.

Meusemann, B.
Reactive sputter etching and reactive ion milling – selectivity, dimensional control, and reduction of MOS-interface degradation.
Journal of Vacuum Science and Technology 16 (6): 1886–1888, Nov.–Dec. 1979.

Mikami, O.; Akiya, H.; Saitoh, T.; Nakagome, H.
CW operation of 1.5 µM GaInAsP/InP buried-heterostructure laser with a reactive ion etched facet.
Electronics Letters 19 (6): 213–215, Mar. 17, 1983.

Miyake, S.
Reactive sputter etching characteristics of Si wafer with CF_4 and O_2 mixed gas.
Japan Society of Precision Engineering Bulletin 15 (2): 89, 1981.

Miyake, S.; Kuwano, H.
Studies on the damage of Si surface caused by reactive sputter etching.
Japan Society of Precision Engineering Journal 47 (12): 1483–1489, Dec. 1981.

Moran, J.M.
High resolution resist patterning using reactive ion etching techniques.
Solid State Technology 24 (4): 195–198, Apr. 1981.

Moran, J.M.; Maydan, D.
High resolution, steep profile resist patterns.
Journal of Vacuum Science and Technology 16 (6): 1620–1624, Nov.–Dec. 1979.

Niggebruegge, U.; Balk, P.
Effect of reactive sputter etching of SiO_2 on the properties of subsequently formed MOS systems.
Solid State Electronics 25 (9): 859–868, Sept. 1982.

Nishimura, T.; Tsukada, T.; Mito, H.
The reactive ion etching for Al films by the CCl_4 and CCl_4/He gas plasmas.
Vacuum Society of Japan Journal 25 (9): 624–630, 1982.

Norstroem, H.; Buchta, R.; Runovic, F.; Wiklund, P.
RIE of SiO_2 in doped and undoped fluorocarbon plasmas.
Vacuum 32 (12): 737–745, Dec. 1982.

Okano, H.; Horiike, Y.
Si etch rate and etch yield with Ar^+/Cl system.
Japanese Journal of Applied Physics 20 (12): 2429–2430, Dec. 1981.

Okano, H.; Yamazaki, T.; Horiike, Y.
High-rate reactive ion etching using a magnetron discharge.
Solid State Technology 25 (4): 166–170, Apr. 1982.

Ono, T.; Ozawa, A.; Yoshihara, H.
Reactive ion etching of polyimide film.
Vacuum Society of Japan Journal 25 (9): 605–612, 1982.

Oshima, M.
Determination of plasma gas temperature during reactive sputter etching.
Japanese Journal of Applied Physics 17 (6): 1157–1158, June 1978.

Oshima, M.
Optical spectroscopy in reactive sputter etching and its application to process control.
Japanese Journal of Applied Physics 20 (4): 683–690, Apr. 1981.

Ozaki, Y.; Hirata, K.
Columnar etching residue generation in reactive sputter etching of SiO_2 and PSG.
Journal of Vacuum Science and Technology 21 (1): 61–65, May-June 1982.

Pang, S.W.; Rathman, D.D.; Silversmith, D.J.; Mountain, R.W.; DeGraff, P.D.
Damage induced in Si by ion milling or reactive ion etching.
Journal of Applied Physics 54 (6): 3272–3277, June 1983.

Paraszczak, J.; Hatzakis, M.
Comparison of CF_4–O_2 and CF_2Cl_2–O_2 plasmas used for the reactive ion etching of single crystal silicon.
Journal of Vacuum Science and Technology 19 (4): 1412–1417, Nov.–Dec. 1981.

Parrens, P.
Anisotropic and selective reactive ion etching of polysilicon using SF_6.
Journal of Vacuum Science and Technology 19 (4): 1403–1407, Nov.–Dec. 1981.

Parrens, P.; Buiguez, F.
Submicron devices fabrication using E-beam masking and RIE.
Microcircuit Engineering '82. International Conference on Microlithography, Grenoble, France, 5–8 Oct. 1982, p. 56–60.

Peccoud, L.
Aluminum etching: Performance comparison between planar and RIE techniques.
Microcircuit Engineering '82. International Conference on Microlithography, Grenoble, France, 5–8 Oct. 1982, p. 221–226.

Pogge, H.B.
Reestablishing parallelism after RIE etching.
IBM Technical Disclosure Bulletin 21 (5): 1849–1850, Oct. 1978.

Pogge, H.B.; Bondur, J.A.; Burkhardt, P.J.
Reactive ion etching of silicon with Cl_2–Ar.
Journal of the Electrochemical Society I25 (11): PC470, 1978.

Ransom, C.M.; Chappell, T.I.; Ephrath, L.M.; Bennett, R.S.
DLTS characterization of RIE processing using Schottky diodes.
Journal of the Electrochemical Society 129 (3): PC105, 1982.

Rathore, H.S.
Method to control the reliability of reactive ion etched vias.
IBM Technical Disclosure Bulletin 24 (3): 1739–1740, Aug. 1981.

Reible, S.A.
Reactive ion etching in the fabrication of niobium tunnel junctions.
IEEE Transactions on Magnetics MAG-17 (1): 303–306, Jan. 1981.

Riseman, J.
Dielectric isolation.
IBM Technical Disclosure Bulletin 25 (8): 4410–4411, Jan. 1983.

Riseman, J.; Shepard, J.F.; Tsang, P.J.
Precision sidewall technology.
IBM Technical Disclosure Bulletin 25 (11B): 6116–6117, Apr. 1983.

Rothman, L.B.; Mauer, J.L.; Schwartz, G.C.; Logan, J.S.
Process of forming tapered vias in SiO_2 by reactive ion etching.
Journal of the Electrochemical Society 127 (3): PC89, 1980.

Rothman, L.B.; Schopen, T.J.; Schwartz, G.C.
Reactive ion etching of organic films.
Journal of Electronic Materials 8 (5): 728, 1979.

Rozich, W.R. Jr.
Reactive etching of silicon nitride.
IBM Technical Disclosure Bulletin 19 (11): 4157, Apr. 1977.

Saia, R.J.; Gorowitz, B.
High rate aluminum etching in a batch loaded reactive ion etcher.
Solid State Technology 26 (4): 247–252, Apr. 1983.

Sato, M.; Nakamura, H.
The effects of mixing N_2 in CCl_4 on aluminum reactive ion etching.
Journal of the Electrochemical Society 129 (11): 2522–2527, Nov. 1982.

Sato, M.; Nakamura, H.
Reactive ion etching of aluminum using $SiCl_4$.
Journal of Vacuum Science and Technology 20 (2): 186–190, 1982.

Schaible, P.M.; Metzger, W.C.; Anderson, J.P.
Reactive ion etching of aluminum and aluminum alloys in an RF plasma containing halogen species.
Journal of Vacuum Science and Technology 15 (2): 334–337, Mar.–Apr. 1978.

Schaible, P.M.; Schwartz, G.C.
Preferential lateral chemical etching in reactive ion etching of aluminum and aluminum alloys.
Journal of Vacuum Science and Technology 16 (2): 377–380, Mar.–Apr. 1979.

Schaible, P.M.; Schwartz, G.C.
Reactive ion etching of silicon.
IBM Technical Disclosure Bulletin 21 (7): 2814–2815, Dec. 1978.

Schaible, P.M.; Schwartz, G.C.
Reactive ion etching of silicon.
IBM Technical Disclosure Bulletin 22 (5): 1819, Oct. 1979.

Schaible, P.M.; Schwartz, G.C.
Temperature measurement during reactive ion etching.
Journal of the Electrochemical Society 127 (3): PC90, 1980.

Schaible, P.M.; Suierveld, J.
RIE of metal composites to form circuitry on ceramics.
IBM Technical Disclosure Bulletin 25 (3A): 971, Aug. 1982.

Schwartz, G.C.; Rothman, L.B.; Schopen, T.J.
Competitive mechanisms in reactive ion etching in a CF_4 plasma.
Journal of the Electrochemical Society 126 (3): 464–469, Mar. 1979.

Schwartz, G.C.; Schaible, P.M.
Reactive ion etching in chlorinated plasmas.
Solid State Technology 23 (11): 85–91, Nov. 1980.

Schwartz, G.C.; Schaible, P.M.
Reactive ion etching of silicon.
Journal of Vacuum Science and Technology 16 (2): 410–413, Mar.–Apr. 1979.

Schwartz, G.C.; Schaible, P.M.
Reactive ion etching of silicon in chlorinated plasmas – parametric study.
Journal of the Electrochemical Society 126 (8): PC374, 1979.

Schwartz, G.C.; Schaible, P.M.
Reactive ion etching of silicon – temperature effects.
Journal of the Electrochemical Society 127 (3): PC89, 1980.

Shibagaki, M.; Horiike, Y.
Si and SiO_2 etching characteristics using reactive ion etching with CF_4–Cl_2 gas mixture.
Japanese Journal of Applied Physics 19 (8): 1579–1580, Aug. 1980.

Shibayama, H.; Kosugi, M.; Hisatsugu, T.
Damage-free dry etching with a triode-type reactive ion etching system.
Fujitsu Scientific and Technical Journal 17 (3): 85–103, Sept. 1981.

Shibayama, H.; Ogawa, T.; Kobayashi, K.; Hisatsugu, T.
Reactive sputter etching system with floating grid.
Japanese Journal of Applied Physics 19 (Suppl. 19-1): 57–60, 1979.

Shibayama, H.; Ogawa, T.; Kobayashi, K.; Kosugi, M.; Hisatsugu, T.
Triode type reactive ion etching system.
Journal of the Electrochemical Society 127 (3): PC108, 1980.

Smolinsky, G.; Truesdale, E.A.; Wang, D.N.K.; Maydan, D.
Reactive ion etching of silicon oxides with ammonia and trifluoromethane: The role of nitrogen in the discharge.
Journal of the Electrochemical Society 129 (5): 1036–1040, May 1982.

Steinbruchel, C.
Langmuir probe measurements on CHF_3 and CF_4 plasmas: The role of ions in the reactive sputter etching of SiO_2 and Si.
Journal of the Electrochemical Society 130 (3): 648–655, Mar. 1983.

Suciu, P.I.; Fuls, E.N.; Boll, H.J.
High-speed NMOS circuits made with X-ray lithography and reactive sputter etching.
IEEE Electron Devices Letters EDL-1 (1): 10–11, Jan. 1980.

Sugano, T.; Asada, K.; Matsumoto, H.
Reactive ion etching by SiF_4 gas.
Annual Report, Engineering Research Institute, Faculty of Engineering, University of Tokyo, vol. 41, p. 105–110, 1982.

Tachi, S.; Miyaki, K.; Tokuyama, T.
Chemical and physical roles of individual reactive ions in Si dry etching.
Japanese Journal of Applied Physics, Part 1, 21 (Suppl. 21-1): 141–146, 1982.

Tsang, P.J.; Shepard, J.F.; Lechaton, J.; Ogura, S.
Characterization of very small (less than 1 µM) sidewall spacers by reactive ion etching.
Journal of the Electrochemical Society 128 (6): PC238, 1981.

Ukai, K.; Hanazawa, K.
Analysis of the imaging accuracy in reactive ion etching.
Journal of Vacuum Science and Technology 15 (2): 338–340, Mar.–Apr. 1978.

Van den Hoek, W.G.M.; Verhaar, R.D.J.
Anisotropic etching of SiO_2 in a parallel plate plasmatherm RIE system.
Microcircuit Engineering '82. International Conference on Microlithography, Grenoble, France, 5–8 Oct. 1982, p. 227–231.

Varshney, R.C.
Self-aligned VMOS structure using reactive ion etching.
IBM Technical Disclosure Bulletin 22 (8B): 3705–3706, Jan. 1980.

Wang, D.N.K.; Maydan, D.; Levinstein, H.J.
Reactive sputter etching and its applications.
Solid State Technology 23 (8): 122–126, Aug. 1980.

Watanabe, I.; Yoshihara, H.; Saitoh, Y.; Matsuo, S.
Very steep profile resist pattern preparation by reactive ion etching with Ar + CH_4 gas mixture.
Japanese Journal of Applied Physics 20 (11): PL804, 1981.

Watanabe, M.O.; Taguchi, M.; Kanzaki, K.; Zohta, Y.
DLTS study of RIE-induced deep levels in Si using P + N diode arrays.
Japanese Journal of Applied Physics, Part 1, 22 (2): 281–286, Feb. 1983.

Watanabe, T.; Shibagaki, M.; Horiike, Y.
Reactive ion etching of poly-Si employing Cl_2 and Cl_2H_2 systems.
Symposium on VLSI Technology, Digest of Papers, Oiso, Japan, 1–3 Sept. 1982, p. 104–105.

Werdt, R.D.; Meeuwissen, W.A.M.
Structuring of bubble overlays by reactive sputter etching in an Ar–H_2O atmosphere.
Journal of Vacuum Science and Technology 16 (6): 2093–2095, 1979.

Whitcomb, E.C.
Selective, anisotropic etching of SiO_2 and PSG in a CHF_3/SF_6, RIE plasma.
Journal of the Electrochemical Society 129 (3): PC104, 1982.

Whitcomb, E.C.; Jones, A.B.
Reactive ion etching of sub-micron $MoSi_2$ poly-Si gates for CMOS SOS devices.
Solid State Technology 25 (4): 121–125, Apr. 1982.

Yabumoto, N.; Oshima, M.; Michikami, O.; Yoshii, S.
Surface damage on Si substrates caused by reactive sputter etching.
Japanese Journal of Applied Physics 20 (5): 893–900, May 1981.

Yamamoto, Y.; Shinada, K.; Itoh, T.; Yada, K.
Damage in the surface region of silicon produced by sputter etching.
Japanese Journal of Applied Physics 13 (3): 551–552, Mar. 1974.

Yeh, J.T.C.; Grege, K.R.
Patterning of poly-para-xylylenes by reactive ion etching.
Journal of Vacuum Science and Technology A 1 (2, Part 1): 604–608, Apr.–June 1983.

Zarowin, C.B.
Plasma etch anisotropy – theory and some verifying experiments relating ion transport, ion energy, and etch profiles.
Journal of the Electrochemical Society 130 (5): 1144–1152, May 1983.

Zarowin, C.B.
Reactive ion etching of gold in a chlorine plasma for microstructure fabrication.
Journal of the Electrochemical Society 125 (8): PC353, 1978.

Zarowin, C.B.
Theory of plasma chemical transport etching of gold in a chlorine plasma.
Thin Solid Films 85 (1): 33–42, 30 Oct. 1981.

Zielinski, L.; Schwartz, G.C.
Reactive ion etching.
Journal of the Electrochemical Society 122 (3): PC71, 1975.

4.4. Ion beam milling and reactive ion beam etching

Anonymous.
Reactive ion beam etching: A progress report.
Solid State Technology 24 (2): 66, Feb. 1981.

Bach, H.
Information covering compact samples and surface layers through thinning and thin-layer chromatography with the ion-beam etching apparatus IEU 100 (Balzers AG).
Mikroskopie 36 (9-10): 317, 1980.

Bach, H.
Method of determining concentration microgradients in glass surface layers by ion beam etching.
Symposium on Mass Transport in Amorphous Solids, Sheffield, Yorks, England, 10–11 Jan. 1973, 4 pp. (Extended Abstracts).

Bach, H.; Baucke, F.G.K.
Investigations of reactions between glasses and gaseous phases by means of photon emission induced during ion-beam etching.
Physics and Chemistry of Glasses 15 (5): 123–129, 1974.

Barr, R.W.; Bazzarre, D.F.
Micro-ion milling of electronic devices with 1-μM periodicity.
Journal of the Electrochemical Society 121 (3): PC110, 1974.

Bazzarre, D.F.; Barr, R.W.
Ion etching of electronic devices with 1-μM periodicity.
Journal of Vacuum Science and Technology 12 (1): 404, Jan.–Feb. 1975.

Betts, R.A.; Pitt, C.W.; Riddle, K.R.; Walpita, L.M.
A comparative study of the dopant profiles in diffused planar optical waveguides by SIMS and guided wave probe.
Applied Physics A (Germany) A31 (1): 29–35, May 1983.

Bollinger, L.D.
Ion beam etching with reactive gases.
Solid State Technology 26 (1): 99–108, Jan. 1983.

Bollinger, L.D.
Ion milling for semiconductor production processes.
Solid State Technology 20 (11): 66–70, Nov. 1977.

Bollinger, L.D.
Production system for ion milling with a large-diameter ion beam.
Journal of Vacuum Science and Technology 15 (2): 789, Mar.–Apr. 1978.

Bollinger, L.D.; Fink, R.C.
A new production technique: Ion milling.
Solid State Technology 23 (11): 79–84, Nov. 1980.

Bollinger, L.D.; Fink, R.C.
A new production technique: Ion milling. 2. Applications.
Solid State Technology 23 (12): 97–103, Dec. 1980.

Bosch, M.A.; Coldren, L.A.; Good, E.
Reactive ion beam etching of InP with Cl_2.
Applied Physics Letters 38 (4): 264–266, 15 Feb. 1981.

Brambley, D.R.; Vanner, K.C.
Role of ion beam etching in magnetic bubble device manufacture.
International Conference on Ion Plating and Allied Technologies (IPAT '79), London, England, July 1979, p. 47–54.

Bresnock, F.J.; Stumpf, Th.
Implementation of adaptive process control to a dry etching process.
Journal of Vacuum Science and Technology 20 (4): 1027–1030, Apr. 1982.

Broadbent, E.K.
Ion beam etching in an evaporator.
Solid State Technology 26 (4): 201–203, Apr. 1983.

Brown, D.M.; Heath, B.A.; Coutumas, T.; Thompson, G.R.
Reactive ion-beam etching of SiO_2 and polycrystalline silicon.
Applied Physics Letters 37 (2): 159–161, 1980.

Castellano, R.N.
Pattern delineation by ion beam milling.
International Conference on Low-Energy Ion Beams, 2nd, Bath, England, 14–17 Apr. 1980, p. 241–245.

Castellano, R.N.; Hokanson, J.L.
A survey of ion beam milling techniques for piezoelectric device fabrication.
Frequency Control Symposium, 29th Annual, Fort Monmouth, New Jersey, 28–30 May, 1975, p. 128–134.

Chapman, R.E.
Redeposition: A factor in ion beam etching topography.
Journal of Materials Science 12 (6): 1125–1133, Jun. 1977.

Chen, C.L.; Wise, K.D.
Gate formation in GaAs MESFET's using ion-beam etching technology.
IEEE Transactions on Electronic Devices ED-29 (10): 1522–1529, Oct. 1982.

Chinn, J.D.; Adesida, I.; Wolf, E.D.
Profile control by chemically-assisted ion beam and reactive ion beam etching.
Applied Physics Letters 43 (2): 185–187, July 15, 1983.

Chinn, J.D.; Fernandez, A.; Adesida, I.; Wolf, E.D.
Chemically assisted ion beam etching of GaAs, Ti, and Mo.
Journal of Vacuum Science and Technology A 1(2, pt. 1): 701–704, Apr.–June 1983.

Coldren, L.A.; Furuya, K.; Miller, B.I.; Rentschler, J.A.
Etched mirror and groove-coupled GaInAsP/InP laser devices for integrated optics.
IEEE Transactions on Microwave Theory and Technology MIT-30 (10): 1667–1676, Oct. 1982.

Cuomo, J.J.; Harper, J.M.E.; Kaufman, H.R.
Small ion sources for large area ion milling.
IBM Technical Disclosure Bulletin 21 (8): 3373–3375, Jan. 1979.

Custode, F.Z.; Fewer, W.; Splinter, M.; Downey, D.F.
VLSI reactive ion beam etching.
Journal of the Electrochemical Society 128 (3): PC105, 1981.

Danilin, B.S.; Kireev, Y.
Ion etching of microstructures in VLSI production.
Mikroelektronika 9 (4): 302–309, Jul.–Aug. 1980.
Transl. in: Soviet Microelectronics 9 (4): 159–165, Jul.–Aug. 1980.

DeGraff, P.D.; Flanders, D.C.
Directional oxygen-ion-beam etching of carbonaceous materials.
Journal of Vacuum Science and Technology 16 (6): 1906–1908, Nov.–Dec. 1979.

Deppe, H.R.; Hasler, B.; Hopfner, J.
Investigations on the damage caused by ion etching of SiO_2 layers at low energy and high dose.
Solid State Electronics 20 (1): 51–55, Jan. 1977.

Dimigin, H.; Hubsch, H.
Generation of deep structures by ion beam etching.
Journal of the Electrochemical Society 121 (3): PC110, 1974.

Dimigin, H.; Luthje, H.
An investigation of ion etching.
Philips Technical Review 35 (7-8): 199–208, 1975.

Dove, D.B.; Molzen, W.; Johnson, G.
Modification to enhance the beam current of a simple ion gun.
Review of Scientific Instruments 47 (3): 299–300, Mar. 1976.

Downey, D.F.; Bottoms, W.R.; Hanley, P.R.
Introduction to reactive ion beam etching.
Solid State Technology 24 (2): 121–127, Feb. 1981.

Dzioba, S.; Naguib, H.M.
Optical spectroscopy during reactive ion beam etching of Si and Al targets.
Journal of Applied Physics 53 (6): 4389–4394, June 1982.

Engemann, J.
Improvements in DC-ion milling of semiconductor devices with small dimensions.
Review of Scientific Instruments 47 (7): 881–883, Jul. 1976.

Fonash, S.J.; Ashok, S.; Singh, R.
Effect of neutral ion beam sputtering and etching on silicon.
Thin Solid Films 90 (3): 231–235, Apr. 23, 1982.

Frank, R.A.; Petvai, S.I.; Schnitzel, R.H.
Ion milling of via holes (IC technology).
IBM Technical Disclosure Bulletin 23 (4): 1409, Sept. 1980.

Furuya, S.; Kobayashi, K.; Yamamoto, S.
Fine pattern fabrication using ion beam etching.
Fujitsu Scientific and Technical Journal 15 (4): 111–120, Dec. 1979.

Gartner, H.M.; Hinderer, J.J.; Petvai, S.I.; Potts, H.R.; Schnitzel, R.H.
Platinum contact/chromium barrier metallurgical processing technique.
IBM Technical Disclosure Bulletin 21 (11): 4503–4504, April 1979.

Geis, M.W.; Lincoln, G.A.; Efremow, N.; Piacentini, W.J.
A novel anisotropic dry etching technique.
Journal of Vacuum Science and Technology 19 (4): 1390–1393, 1981.

Gerlachmeyer, U.; Coburn, J.W.; Kay, E.
A method for increasing the etch rate ratio of oxides to non-oxides in inert-gas ion milling processes.
Journal of Applied Physics 51 (6): 3362–3364, Jun. 1980.

Gildenblatt, G.; Heath, B.A.; Katz, W.
Interface states induced in silicon by tungsten as a result of reactive ion beam etching.
Journal of Applied Physics 54 (4): 1855–1859, Apr. 1983.

Gloersen, P.G.
Ion-beam etching.
Journal of Vacuum Science and Technology 12 (1): 28–35, Jan.–Feb. 1975.

Gloersen, P.G.
Masking for ion-beam etching.
Solid State Technology 19 (4): 68–73, Apr. 1976.

Gokan, H.; Esho, S.
Fine pattern fabrication by ion beam etching.
Journal of the Vacuum Society of Japan 20 (11): 386–393, 1977.

Gokan, H.; Esho, S.
Gap reduction technique for obtaining submicron geometries utilizing redeposition effect.
Journal of Vacuum Science and Technology 19 (1): 32–35, May–Jun. 1981.

Gokan, H.; Esho, S.
Pattern fabrication by oblique incidence ion-beam etching.
Journal of Vacuum Science and Technology 18 (1): 23–27, Jan.–Feb. 1981.

Gokan, H.; Esho, S.
Pattern profile control utilizing shadow effect in oblique ion-beam etching.
Journal of Vacuum Science and Technology 19 (1): 28–31, May–Jun. 1981.

Hakhu, J.K.
Reactive ion beam etching for VLSI.
Symposium on VLSI Technology, Maui, Hawaii, 9–11 Sept. 1981, p. 66–67.

Harper, J.M.E.
Effects of beam, target and substrate potentials in ion beam processing.
Thin Solid Films 92 (1/2): 107–114, June 11, 1982.

Harper, J.M.E.; Cuomo, J.J.; Kaufman, H.R.
Developments in broad-beam ion source technology and applications.
American Vacuum Society, 29th National Symposium, Baltimore, Maryland, 16–19 Nov. 1982, p. 337–339.

Harper, J.M.E.; Cuomo, J.J.; Kaufman, H.R.
Technology and applications of broad-beam ion sources used in sputtering–2. Applications.
Journal of Vacuum Science and Technology 21 (3): 737–756, Sept.–Oct. 1982.

Harper, J.M.E.; Cuomo, J.J.; Leary, P.A.; Summa, G.M.; Kaufman, H.R.; Bresnock, F.J.
Low energy ion beam etching.
Journal of the Electrochemical Society 128 (5): 1077–1083, May 1981.

Harper, J.M.E.; Kaufman, H.R.
Technique for improved selectivity in ion beam etching.
IBM Technical Disclosure Bulletin 23 (5): 2143–2145, Oct. 1980.

Hawkins, D.T.
Ion milling (ion beam etching), 1975–1978: A bibliography.
Journal of Vacuum Science and Technology 16 (4): 1051–1071, July–Aug. 1979.

Hayashi, T.; Miyamura, M.; Komiya, S.
Observations of C_mF_n radicals in reactive ion beam etching.
Japanese Journal of Applied Physics 21 (12, pt. 2): 755–757, Dec. 1982.

Heath, B.A.
Etching SiO_2 in a reactive ion beam.
Solid State Technology 24 (10): 75–78, Oct. 1981.

Heath, B.A.
Selective reactive ion-beam etching of SiO_2 over polycrystalline Si.
Journal of the Electrochemical Society, 129 (2): 396–402, 1982.

Herdzik, R.J.; Sarkary, H.G.; Totta, P.A.
Ion milling to remove halo.
IBM Technical Disclosure Bulletin 23 (11): 4920, Apr. 1981.

Herrmann, W.C. Jr.; McNeil, J.R.
Ion beam milling as a diagnostic for optical coatings.
Applied Optics 20 (11): 1899–1901, 1 Jun., 1981.

Hiraoka, H.
Ion beam etchings using HCl.
IBM Technical Disclosure Bulletin 22 (3): 773, July 1979.

Hoffmann, M.; Reimer, L.
Channeling contrast on metal surfaces after ion-beam etching.
Scanning 4 (2): 91–93, 1981.

Hosaka, S.; Hashimoto, S.
Influence of sample inclination and rotation during ion-beam etching on ion-etched structures.
Journal of Vacuum Science and Technology 15 (5): 1712–1717, Sept.–Oct. 1978.

Jin, W.; Meng, X.; You, D.; Xu, X.
Reactive ion beam etching and its application.
Chinese Journal of Semiconductors 4 (1): 97–100, Jan. 1983.

Jolly, T.W.; Clampitt, R.
Ion milling – the competing technology.
Semiconductor International '82: 64–67, 1982.

Kegel, B.; Liebel, G.
Micro ion milling techniques reviewed for producing high resolution devices for surface wave acoustics, magnetic domain memories, field transistors and integrated optics.
International Vacuum Congress, 7th, and International Conference on Solid Surfaces, 3rd, Vienna, Austria, 12–16 Sept. 1977, p. 1453–1455.

Kiddon, J.; Sellers, J.; Coutumas, T.; Brewer, T.
Reactive ion milling of aluminum and alloys.
Journal of the Electrochemical Society 128 (3): PC105, 1981.

Kireev, V.Y.; Kuznetsova, V.V.; Lavrischev, V.P.; Makhov, V.I.; Yastrebov, V.B.
Production of submicron-size elements in films of various materials by reactive ion beam etching.
Soviet Microelectronics 10 (6): 237–242, Nov.–Dec. 1981.

Kleinsasser, A.W.; Buhrman, R.A.
Fabrication of Josephson tunnel junctions by reactive ion milling.
Journal of the Electrochemical Society 127 (3): PC109, 1980.

Krongelb, S.
Control of edge build-up in sputter etching and ion milling.
IBM Technical Disclosure Bulletin 21 (3): 1253–1255, Aug. 1978.

Krumme, J.-P.; Dimigen, H.
Ion-beam etching of groove patterns into garnet films.
IEEE Transactions on Magnetics MAG-9 (3): 405–408, Sept. 1973.

Laznovsky, W.
Advances in low-energy ion beam technology.
Research/Development 26 (8): 47–48, 50, 52, 54–55, Aug. 1975.

Lee, R.E.
Ion beam etching.
Electronic Production 9 (6): 41–45, 49–50, June 1980.

Lee, R.E.
Microfabrication by ion beam etching.
Journal of Vacuum Science and Technology 16 (2): 164–170, Mar.–Apr. 1979.

Lee, R.E.
Microfabrication by ion beam etching.
Semiconductor International 3 (1): 73–80, 82, Jan. 1980.

Lincoln, G.A.; Geis, M.W.; Mahoney, L.J.; Chu, A.; Vojak, B.A.; Nichols, K.B.; Piacentini, W.J.; Efremow, N.; Lindley, W.T.
Ion beam assisted etching for GaAs device applications.
Journal of Vacuum Science and Technology 20 (3): 786–789, Mar. 1982.

Mader, L.; Höpfner, J.
Ion beam etching of silicon dioxide on silicon.
Journal of the Electrochemical Society 123 (12): 1893–1898, Dec. 1976.

Mader, L.; Widmann, D.; Höpfner, J.
Ion beam etching of silicon dioxide layers for MOS devices.
Electrochemical Society Spring Meeting (Extended Abstracts), Toronto, Canada, 11–16 May 1975, p. 432–433.

Matsinger, J.H.; Sandy, F.
A microprocessor-controlled ion machining apparatus.
IEEE Transactions on Industrial Electronics and Control Instrumentation IECI-22 (3): 295–300, Aug. 1975.

Matsui, S.; Mizuki, S.Y.; Yamato, T.; Aritome, H.; Namba, A.
Reactive ion beam etching of silicon carbide.
Japanese Journal of Applied Physics 20 (1): 38–40, Jan. 1981.

Matsui, S.; Morikawi, K.; Masuda, N.; Nakamura, T.; Aritome, H.; Namba, S.
Fabrication of SiO_2 grating patterns with vertical sidewalls by SOR X-ray lithography and reactive ion-beam etching.
Japanese Journal of Applied Physics 20 (9): 1735–1740, Sept. 1981.

Matsui, S.; Yamato, T.; Aritome, H.; Namba, S.
Microfabrication of $LiNbO_3$ by reactive ion beam etching.
Japanese Journal of Applied Physics 19 (8): 463–465, Aug. 1980.

Matsuo, S.; Adachi, Y.
Reactive ion-beam etching using a broad beam ECR ion source.
Japanese Journal of Applied Physics Part 2, 21 (1): 4–6, Jan. 1982.

Mayer, T.M.; Barker, R.A.
Reactive ion beam etching with CF_4 – characterization of a Kaufman ion source and details of SiO_2 etching.
Journal of the Electrochemical Society 129 (3): 585–591, Mar. 1982.

Mayer, T.M.; Barker, R.A.; Whitman, L.J.
Investigation of plasma etching mechanisms using beams of reactive gas ions.
Journal of Vacuum Science and Technology 18 (2): 349–352, Mar. 1981.

Melliar-Smith, C.M.
Ion etching for pattern delineation.
Journal of Vacuum Science and Technology 13 (5): Sept.–Oct. 1976.

Miyamura, M.; Tsukakoshi, O.; Komiya, S.
A 26-cm electron-cyclotron-resonance ion source for reactive ion beam etching of SiO_2 and Si.
Journal of Vacuum Science and Technology 20 (4): 986–988, Apr. 1982.

Monk, G.W.; Thompson, G.R.
Ion beam etching equipment for production of bubble memory devices.
Journal of Applied Physics 50 (3, Part 2): 2299, Mar. 1979.

Moriwaki, K.; Aritome, H.; Namba, S.
Diffraction gratings for X-ray fabricated by reactive ion beam etching.
Journal of the Electrochemical Society 129 (3): PC112, 1982.

Moriwaki, K.; Aritome, H.; Namba, S.
Fabrication of 80 nm-wide lines in FPM resist by H^+ beam exposure.
Japanese Journal of Applied Physics 20 (12): L881–884, Dec. 1981.

Mundy, J.N.; Rothman, S.J.
An apparatus for microsectioning diffusion samples by sputtering.
Journal of Vacuum Science and Technology A 1 (1): 74–76, Jan.–Mar. 1983.

Neureuther, A.R.; Liu, C.Y.; Ting, C.H.
Modeling ion milling.
Journal of Vacuum Science and Technology 16 (6): 1767–1771, Nov.–Dec. 1979.

Okano, H.; Horike, Y.
Reactive ion beam etching of SiO_2 and poly-Si employing C_2F_6, SiF_4 and BF_3 gases.
Japanese Journal of Applied Physics Part 1 21 (5): 696–701, May 1982.

Pawar, P.G.; Duhamel, P.; Monk, G.W.
Effects of ion beam milling on surface topography.
Electron Microscopy Society of America, 31st Annual Meeting (Extended Abstracts), New Orleans, Lousiana, 14–17 Aug. 1973, p. 84–85.

Pelegrini, F.; Teale, R.W.; Abell, S.
Electron spin resonance of single crystals of terbium doped $GdAl_2$ in the ferromagnetic and paramagnetic phases.
Journal of Magnetism and Magnetic Materials 29 (1-3): 105–112, Oct. 1982.

Petvai, S.I.; Schnitzel, R.H.; Frank, R.
Cleaning of vias by ion milling (LSI devices).
Thin Solid Films 53 (1): 111–116, 15 Aug. 1978.

Pitt, C.W.; Singh, S.P.
Neutralized ion beam milling: Anomalous sputter yield behaviour.
Electronics Letters 16 (19): 721–722, 11 Sept. 1980.

Powell, R.A.
Reactive ion beam etching of GaAs in CCl_4.
Japanese Journal of Applied Physics Part 2 21 (3): 170–172, Mar. 1982.

Powell, R.A.
Reactive ion beam etching of $MoSi_2$ in CCl_4.
Journal of the Electrochemical Society 130 (5): 1164–1167, May 1983.

Rangelow, I.W.
Computer simulation of line edge profiles undergoing ion bombardment.
Journal of Vacuum Science and Technology A 1 (2, pt. 1): 410–414, Apr.–June 1983.

Rangelow, I.W.; Radzimski, Z.; Czarczynski, W.
Ion milling source with slot extraction system.
Journal of Vacuum Science and Technology A 1 (2, pt. 1): 244–247, Apr.–June 1983.

Revell, P.J.; Evans, A.C.
Ion-beam etching using saddle field sources.
Thin Solid Films 86 (2-3): 117–123, 1981.

Revell, P.J.; Goldspink, G.F.
Reactive ion beam etching of silicon compounds with a saddle field ion source.
Journal of Vacuum Science and Technology 19 (4): 1398–1402, Nov.–Dec. 1981.

Robertson, D.D.
Advances in ion beam milling (semiconductor device manufacture).
Solid State Technology 21 (1): 57–60, Dec. 1978.

Robinson, R.S.
Thirty centimeter diameter ion milling source.
Journal of Vacuum Science and Technology 15 (2): 277–280, Mar.–Apr. 1978.

Rost, M.; Erler, H.J.; Giegenga, H.; Fiedler, O.; Weissman, C.
Ion-beam etching and sputtering of polytetrafluoroethylene (PTFE).
Thin Solid Films 20 (1): PS15, 1974.

Schinke, D.P.; Smith, R.G.; Spencer, E.G.; Galvin, M.F.
Thin-film distributed-feedback laser fabricated by ion milling.
Applied Physics Letters 21 (10): 494–496, 15 Nov. 1972.

Singer, I.L.
Surface morphology produced by ion milling on ion-implanted 18Cr8Ni steels.
Journal of Vacuum Science and Technology 18 (2): 175–178, Mar. 1981.

Singer, I.L.; Murday, J.S.; Comas, J.
Preferential sputtering from disordered GaAs.
Journal of Vacuum Science and Technology 18 (2): 161–163, Mar. 1981.

Singh, R.; Fonash, S.J.; Ashok, S.; Caplan, P.J.; Shappirio, J.; Hage-Ali, M.; Ponpon, J.
Electrical, structural, and bonding changes induced in silicon by H, Ar, and Kr ion-beam etching.
Journal of Vacuum Science and Technology A 1 (2, pt. 1): 334–336, Apr.–June 1983.

Smith, D.L.
Dry etching for precise definition of integrated circuit patterns.
Perkin–Elmer Technical News 10 (1): 22–30, June 1982.

Smith, H.I.
Ion beam etching.
Electrochemical Society Spring Meeting (Extended Abstracts), Washington, D.C., 2–7 May 1976, p. 148.

Smith, H.I.
Ion beam etching of surface gratings.
IEEE Transactions on Sonics and Ultrasonics VSU-21 (1): 77, 1974.

Smith, H.I.; Melngailis, J.; Williamson, R.C.; Brogan, W.T.
Ion beam etching of surface gratings (for acoustic surface wave filters).
Journal of Vacuum Science and Technology 10 (6): 1127, Nov.–Dec. 1973.

Smith, R.; Makh, S.S.; Walls, J.M.
Surface morphology during ion etching. The influence of redeposition.
Philosophical Magazine A 47 (4): 453–481, Apr. 1983.

Spencer, E.G.; Schmidt, P.H.
Ion-beam techniques for device fabrication.
Journal of Vacuum Science and Technology 8 (5): S52–70, Sept.–Oct. 1971.

Spencer, E.G.; Schmidt, P.H.; Fischer, R.F.
Microstructure arrays produced by ion milling.
Applied Physics Letters 17 (8): 328–332, 15 Oct. 1970.

Springer, S.
Ion beam etching: A method for the preparation of thin foils.
Praktische Metallographie 11: 311–322, June 1974.

Tachi, S.; Miyaki, K.; Tokuyama, T.
Chemical and physical sputtering in F^+ ion-beam etching of Si.
Japanese Journal of Applied Physics 20 (6): PL411, 1981.

Taubenblatt, M.A.; Helms, Cr.R.
Ion knock-on broadening effects in Auger sputter profiling studies of ultra-thin SiO_2 layers on Si.
Journal of Applied Physics 54 (5): 2667–2671, May 1983.

Taylor, N.; Johannessen, J.S.; Spicer, W.E.
Crater-edge profiling in interface analysis employing ion-beam etching and AES.
Applied Physics Letters 29 (8): 497–499, Oct. 15, 1976.

Tsuge, H.; Esho, S.; Gokan, H.
Simulation of ion-beam etched pattern profiles.
Journal of Vacuum Science and Technology 19 (2): 221–224, July–Aug. 1981.

Venkatesan, T.; Taylor, G.N.; Wagner, A.; Wilkens, B.; Barr, D.
Plasma-developed ion implanted resists with submicron resolution.
Journal of Vacuum Science and Technology 19 (4): 1379–1384, Nov/Dec. 1981.

Weissmantel, C.; Rost, M.; Fiedler, O.; Erler, H.J.; Geigengack, H.; Horn, J.
Interaction of ion beams with polymer surfaces leading to etching and sputtering processes.
International Vacuum Congress, 6th, Kyoto, Japan, 25–29 Mar. 1974, p. 439–442.

Yamane, Y.; Yamasaki, K.; Mizutani, T.
Annealing behavior of damage introduced in GaAs by reactive ion beam etching.
Japanese Journal of Applied Physics, Pt. 2 21 (9): 537–538, Sept. 1982.

Yano, H.; Hashimoto, H.; Toyama, Y.
Damage caused by Ar ion-beam etching.
Journal of the Electrochemical Society 125 (3): PC155, 1978.

Yasuda, H.
Application of ion milling techniques to microparabolic surface formation.
Journal of Applied Physics 45 (1): 484–486, Jan. 1974.

Yasuda, H.
Ion milling by beam scanning techniques – effect of beam diameter on milled surface shape.
Japanese Journal of Applied Physics (Suppl. 2, Part 1): 431–434, 1974.

Yasuda, H.
Surface formation.
Journal of Applied Physics 45 (1): 484–486, Jan. 1974.

Youngner, D.W.; Haynes, C.M.
Modeling ion beam milling.
Journal of Vacuum Science and Technology 21 (2): 677–680, July–Aug. 1982.

4.5. End-point detection and plasma diagnostics

Bayraktaroglu, B.; Johnson, R.L.
Silicon nitride–gallium arsenide MIS structures produced by plasma enhanced deposition.
Journal of Applied Physics 52 (5): 3515–3519, 1981.

Bunyard, G.B.; Raby, B.A.
Plasma process development and monitoring via mass spectrometry.
Solid State Technology 20 (12): 53–57, Dec. 1977.

Busta, H.H.
End point detection with laser interferometry (in plasma etching).
SPIE Proceedings 276: 164–169, 1981.

Busta, H.H.; Lajos, R.E.
Ellipsometric end point detection during plasma etching.
International Electron Devices Meeting, Washington, D.C., 5–7 Dec. 1977, p. 12–15.

Chen, M.; Coburn, J.W.
Applying emission spectroscopy to the scale-up problem in plasma processing.
IBM Technical Disclosure Bulletin 22 (12): 5431, May 1980.

Collins, J.C.; Pavone, P.J.
Automatic process end-point detection system.
IBM Technical Disclosure Bulletin 27 (5): 1342–1343, Oct. 1974.

Curtis, B.J.
Optical end-point detection for the plasma etching of aluminum.
Solid State Technology 23 (4): 129–132, Apr. 1980.

Curtis, B.J.; Brunner, H.J.
End point determination of aluminum CCl_4 plasma etching by optical emission spectroscopy.
Journal of the Electrochemical Society 125 (5): 829–830, May 1978.

Danner, D.A.; Flamm, D.L.; Mucha, J.A.
Downstream atomic monitoring for absolute etch rate determinations.
Journal of the Electrochemical Society 130 (4): 905–907, Apr. 1983.

Dennison, R.W.
Mass spectrometry applied to a reactive ion mill.
Solid State Technology 23 (9): 117–120, Sept. 1980.

Eisele, K.M.; Hofman, D.
Application of plasma mass spectrometry to plasma etch processes.
International Vacuum Congress, 8th, Cannes, France, 22–26 Sept. 1980, v. 1, p. 62–65.

Frieser, R.G.; Nogay, J.
Optical spectroscopy applied to the study of plasma etching.
Applied Spectroscopy 34 (1): 31–33, Jan.–Feb. 1980.

Geipel, H.J.
End-point detection for reactive ion etching.
IBM Technical Disclosure Bulletin 20 (2): 541–542, July 1977.

Gourrier, S.
Real time study of plasma/surface reactions by optical means.
Acta Electronica 24 (3): 229–238, 1981–1982.

Greene, J.E.
Optical spectroscopy for diagnostics and process control during glow discharge etching and sputter deposition.
Journal of Vacuum Science and Technology 15 (5): 1718–1729, Sept.–Oct. 1978.

Griffiths, D.P.; Bradley, S.H.
Plasma etching as a diagnostic technique in silicon surface studies.
Journal of Materials Sceince 12 (5): 1019–1027, May 1977.

Harshbarger, W.R.; Norton, P.; Porter, R.A.
Optical detector to monitor plasma etching.
Journal of Electronic Materials 6 (6): 739, 1977.

Harshbarger, W.R.; Porter, R.A.; Miller, T.A.; Norton, P.
A study of the optical emission from an RF plasma during semiconductor etching.
Applied Spectroscopy 31 (3): 201–207, May–June 1977.

Harshbarger, W.R.; Porter, R.A.; Norton, P.
Optical detector to monitor plasma etching.
Journal of Electronic Materials 7 (3): 429–440, May 1978.

Haynes, C.V.; Swanson, J.G.; Skinner, D.K.
Preparation and properties of plasma-anodized alumina InP interfaces using in situ end point detection methods.
Thin Solid Films 103 (1-2): 77–93, May 13, 1983.

Hosaka, S.; Sakudo, N.; Hashimoto, S.
Monitoring secondary ions during ion etching.
Journal of Vacuum Science and Technology 16 (3): 913–917, May–June 1979.

Hirobe, K.; Tsuchimoto, T.
End-point detectability in plasma etching.
Journal of the Electrochemical Society 125 (3): PC146, 1978.

Hirobe, K.; Tsuchimoto, T.
End-point detection in plasma etching by optical emission spectroscopy.
Journal of the Electrochemical Society 127 (1): 234–235, Jan. 1980.

Hitchman, M.L.; Eichenberger, V.
A simple method of end-point determination for plasma etching.
Journal of Vacuum Science and Technology 17 (6): 1378–1381, Nov.–Dec. 1980.

Johnson, D.
Optical methods detect end point in plasma etching.
Industrial Research and Development 22 (10): 181, 183–184, 186, Oct. 1980.

Kay, E.; Coburn, J.W.; Kruppa, G.
Mass spectrometric studies of both polymerizing and etching fluorocarbon glow discharges.
Vide 31 (183): 89–96, May–June 1976.

Khoury, H.A.; Tomkins, B.E.
Front wafer registration device for batch process etch end point detection system.
IBM Technical Disclosure Bulletin 20 (5): 1756–1759, Oct. 1977.

Kleinknecht, H.P.; Meier, H.
Optical monitoring of the etching of SiO_2 and Si_3N_4 on Si by the use of grating test patterns.
Journal of the Electrochemical Society 125 (5): 798–803, May 1978.

Kolodner, P.; Katzir, A.; Hartsough, N.
End point detection and etch-rate measurement during reactive ion etching using fluorescent polymer films.
Journal of Vacuum Science and Technology B 1 (2): 501–504, Apr.–June 1983.

Leahy, M.F.
Endpoint detection of etching reactions: Optical monitoring of polysilicon, silicon nitride, and aluminum etch processes.
Microelectronics Measurement Technology Seminar, 3rd Annual, San Jose, California, 17–18 March 1981, p. 1–20.

Lehmann, H.W.; Heeb, E.; Frick, K.
Plasma diagnostics by time resolved mass spectrometry.
Solid State Technology 24 (10): 69–74, Oct. 1981.

Lewin, T.P.
A unique technique for simultaneous multiple wafer etching with individual end-point detection.
Semiconductor International '82, p. 191–198, 1982.

Marcoux, P.J.; Pang, D.F.
Methods of end point detection for plasma etching.
Solid State Technology 24 (4): 115–122, Apr. 1981.

Marcoux, P.J.; Pang, D.F.
Optical methods for end point detection in plasma etching.
SPIE Proceedings, 276: 170–177, 1981.

Millard, M.; Kay, E.
Difluorocarbene emission spectra from fluorocarbon plasmas and its relationship to fluorocarbon polymer formation.
Journal of Vacuum Science and Technology 18 (2): 343–344, Mar. 1981.

Nishizawa, J.; Hayasaka, N.
In situ observation of plasmas for dry etching by IR spectroscopy and probe methods.
Thin Solid Films 92 (1/2): 189–198, June 11, 1982.

Oshima, M.
Monitoring of dry etching process of SiO_2 on Si by using mass spectra.
Japanese Journal of Applied Physics 17 (3): 579–580, Mar. 1978.

Oshima, M.
Use of mass spectra for end point detection in etching SiO_2 films on Si.
Japanese Journal of Applied Physics 20 (7): 1255–1263, 1981.

Raby, B.A.
Mass spectrometric study of plasma etching.
Journal of Vacuum Science and Technology 15 (2): 205–208, Mar.–Apr. 1978.

Stafford, B.B.; Gorin, G.J.
Optical emission end point detecting for monitoring oxygen plasma photoresist stripping.
Solid State Technology 20 (9): 51–55, Sept. 1977.

Sternheim, M.; Van Gelder, W.; Hartman, A.W.
A laser interferometer system to monitor dry etching of patterned silicon.
Journal of the Electrochemical Society 130 (3): 655–658, Mar. 1983.

Tsukada, T.; Ukai, K.
End point determination for plasma etching with a double beam optical-emission spectrometer.
Journal of the Electrochemical Society 127 (3): PC91, 1980.

Ukai, K.; Hanazawa, K.
End point determination of aluminum reactive ion etching by discharge impedance monitoring.
Journal of Vacuum Science and Technology 16 (2): 385–387, Mar.–Apr. 1979.

Viswanathan, N.S.
Ion sampling from RF discharges at the cathode.
IBM Technical Disclosure Bulletin 20 (4): 1409–1410, Sept. 1977.

Wang, C.W.; Gelernt, B.
Optical end-point detection for plasma etching of thermal oxide and phosphosilicate glass.
Microelectronics Measurement Technology Seminar, 3rd Annual, San Jose, California, 17–18 March 1981, p. 43–62.

Zijlstra, P.A.; Beenakker, C.I.M.
Evaluation of silicon-chemiluminescence monitoring as a novel method for atomic fluorine determination and end point detection in plasma etch systems.
Applied Spectroscopy 35 (4): 413–418, July–Aug. 1981.

4.6. Resists and masking materials

Aboelfotoh, M.O.; Masud, C.; Polcari, M.R.
Process for metal patterning using silicon layer mask defined by reactive ion etching photoresist pattern.
IBM Technical Disclosure Bulletin 25 (6): 2762–2763, Nov. 1982.

Adesida, I.; Chinn, J.D.; Rathbun, L.; Wolf, E.D.
Dry development of ion beam exposed PMMA resist.
Journal of Vacuum Science and Technology 21 (2): 666–671, July–Aug. 1982.

Ahn, K.Y.; Schwartz, G.C.
Fabrication of mask for dry etching of microcircuits.
IBM Technical Disclosure Bulletin 21 (4): 1713–1714, Sept. 1978.

Asakawa, H.; Kogure, O.
A highly sensitive positive electron resist (FBM-G).
1982 Symposium on VLSI Technology, Digest of Papers, Oiso, Japan, 1–3 Sept. 1982, p. 88–89. Published by IEEE, New York.

Asmussen, F.; Betz, H.; Chen, J.-T.; Heuberger, A.; Pongratz, S.; Sotobayashi, H.; Schnabel, W.
Properties of cross-linked positive acting X-ray resists fabricated on the basis of poly(methylmethacrylate-co-methacrylyl chloride).
Journal of the Electrochemical Society 130 (1): 180–184, Jan. 1983.

Badami, D.A.; Bergendahl, A.S.; Hakey, M.C.
Method of characterizing ion-implanted photoresist.
IBM Technical Disclosure Bulletin 25 (12): 6546, May 1983.

Bassous, E.; Ephrath, L.M.; Pepper, G.; Mikalsen, D.J.
A three-layer resist system for deep UV and RIE microlithography on nonplanar surfaces.
Journal of the Electrochemical Society 130 (2): 478–484, Feb. 1983.

Cantagrel, M.
Comparison of the properties of different materials used as masks for ion-beam etching.
Journal of Vacuum Science and Technology 12 (6): 1340–1343, Nov.–Dec. 1976.

Chang, M.S.; Chen, J.T.
Plasma etching of inorganic resists.
Journal of Electronic Materials 8 (5): 727, 1979.

Charlet, B.; Peccoud, L.
Fast plasma hardening of microwave resist layers.
Microcircuit Engineering '82. International Conference on Microlithography, Grenoble, France, 5–8 Oct. 1982, p. 215–221.

Cox, D.E.
Reducing the thickness of resist milling masks.
IBM Technical Disclosure Bulletin 21 (8): 3406–3408, Jan. 1979.

Cox, D.E.; McGouey, R.P.
Minimal metal mask for reactive ion etching polyimide.
IBM Technical Disclosure Bulletin 23 (2): 830, July 1980.

Ephrath, L.M.
Teflon polymer mask for RIE of contact holes.
IBM Technical Disclosure Bulletin 25 (9): 4587–4588, Feb. 1983.

Fukuda, M.
CMS gives impact on dry etching process in VLSI production.
JEE 19 (188): 40–43, Aug. 1982.

Galicki, A.
Stripping of resist in situ in plasma etching chamber.
IBM Technical Disclosure Bulletin 21 (1): 128–129, June 1978.

Geis, M.W.; Randall, J.N.; Deutsch, T.F.; DeGraff, P.D.; Krohn, K.E.; Stern, L.A.
Self-developing resist with submicrometer resolution and processing stability.
Applied Physics Letters 43 (1): 74–76, July 1, 1983.

Gipstein, E.; Gritter, R.J.; Need, O.U. III; Yoon, D.Y.
Radiation sensitive, high temperature, RIE resistant polymeric resist.
IBM Technical Disclosure Bulletin 20 (3): 1205, Aug. 1977.

Gloersen, P.G.
Masking for ion-beam etching.
Solid State Technology 19 (4): 68–73, Apr. 1976.

Grabbe, P.; Howard, R.E.; Hu, E.L.; Tennant, D.M.
Metal-on-polymer masks for reactive ion etching.
Journal of the Electrochemical Society 129 (3): PC112, 1982.

Gregor, L.V.; Kaplan, L.H.; Lane, J.G.; Mathad, S.G.; Moreau, W.M.; Zingerman, J.R.
Tri-level resist structure for low-dosage high-resolution E-beam lithography.
IBM Technical Disclosure Bulletin 24 (7B): 3837–3838, Dec. 1981.

Greschner, J.; Trumpp, H.J.
Lift-off process for producing a dense metallization pattern.
IBM Technical Disclosure Bulletin 25 (8): 4481–4482, Jan. 1983.

Harada, K.
Additives that improve positive resist durability for plasma etching.
Journal of the Electrochemical Society 127 (2): 491–497, Feb. 1980.

Harada, K.
Dry etching durability of positive electron resists.
Journal of Applied Polymer Science 26 (10): 3395–3408, Oct. 1981.

Harada, K.
Plasma etching durability of poly(methyl methacrylate).
Journal of Applied Polymer Science 26 (6): 1961–1973, June 1981.

Harada, K.; Kogure, O.; Murase, K.
Poly (phenyl methacrylate-co-methacrylic acid) as a dry-etching durable positive electron resist.
IEEE Transactions on Electron Devices ED-29 (4): 518–524, Apr. 1982.

Harada, K.; Suguwara, S.
Dry-etching durabilities of positive electron resists.
Electrical Communications Laboratory Technical Journal 30 (12): 2919–2932, 1981.

Helbert, J.N.; Schmidt, M.A.
Effect of composition on resist dry-etching: Susceptibility of vinyl polymers and photoresists.
IEEE Transactions on Electron Devices, ED-29 (4): 518–524, 1982.

Hieke, E.; Beinvogel, W.
Trimming of negative electron resist for multilevel metallization.
Microcircuit Engineering '82. International Conference on Microlithography, Grenoble, France, 5–8 Oct. 1982, p. 338–340.

Hiraoka, H.; Pacansky, J.
UV hardening of photo- and electron-beam resist patterns.
Journal of Vacuum Science and Technology 19 (4): 1131–1135, Nov/Dec. 1981.

Hoeg, A.J. Jr.; Kroll, C.T.; Stephens, G.B.
Metal lift-off process with a self-aligned insulation planarization.
IBM Technical Disclosure Bulletin 24 (9): 4839–4840, Feb. 1982.

Horiike, Y.; Sugawara, T.; Okano, H.; Shibagaki, M.; Ueda, Y.
Cl_2/Ar plasma etching of a contaminated layer on Si induced by fluorocarbon gas plasma.
Japanese Journal of Applied Physics 20 (4): 803–804, Apr. 1981.

Horwitz, C.M.
New dry etch for Al and Al–Cu–Si alloy: Reactively masked sputter etching with SiF_4.
Applied Physics Letters 42 (10): 898–900, May 15, 1983.

Hu, E.L.; Howard, R.E.; Grabbe, P.; Tennant, D.M.
Ion beam processing using metal on polymer masks.
Journal of the Electrochemical Society 130 (5): 1171–1173, May 1983.

Hu, E.L.; Tennant, D.M.; Howard, R.E.; Jackel, L.D.; Grabbe, P.
Vertical silicon membrane arrays patterned with tri-level E-beam resist.
Journal of Electronic Materials 11 (5): 883–888, Sept. 1982.

Huggett, P.G.; Frick, K.; Lehmann, H.W.
Development of silver sensitized germanium selenide photoresist by reactive sputter etching in SF_6.
Applied Physics Letters 42 (7): 592–594, Apr. 1, 1983.

Hughes, H.G.; Goodner, W.R.; Wood, T.E.; Smith, J.N.; Keller, J.V.
Photoresist development by plasma.
Polymer Engineering Science 20 (16): 1093–1096, Mid-Nov. 1980.

Hunt, B.D.; Buhrman, R.A.
Multilayer, high resolution, ion-bombardment-tolerant electron resist system.
Journal of Vacuum Science and Technology 19 (4): 1308–1312, Nov/Dec. 1981.

Hunter, W.R.; Holloway, T.C.; Chatterjee, P.K.; Tasch, A.F. Jr.
A new edge-defined approach for submicrometer MOSFET fabrication.
IEEE Electron Device Letters EDL-2 (1): 4–6, Jan. 1981.

Imamura, S.
Chloromethylated polystyrene as a dry etching-resistant negative resist for submicron technology.
Journal of the Electrochemical Society 126 (9): 1628–1630, Sept. 1979.

Johnson, C.; Patnaik, B.
Polyimide mask for reactive ion etching of metal lines.
IBM Technical Disclosure Bulletin 22 (4): 1446–1448, Sept. 1979.

Johnson, L.F.; Ingersoll, K.A.; Dalton, J.V.
Planarizing of phosphosilicate glass films on patterned silicon wafers.
Journal of Vacuum Science and Technology B 1 (2): 487–489, Apr.–June 1983.

Kadota, K.; Taki, Y.; Shimizu, S.
New positive photoresist for critical dimension control.
SPIE Proceedings 275: 173–181, 1981.

Karapiperis, L.; Lee, C.A.
Ion beam fabrication of 400 AA, high aspect-ratio lines in poly methyl methacrylate (PMMA).
Journal of Vacuum Science and Technology 16 (6): 1625, Nov.–Dec. 1979.

Kuwano, H.; Yoshida, K.; Yamazaki, S.
Dry development of resists exposed to focused gallium ion beam (semiconductor fabrication).
Japanese Journal of Applied Physics 19 (10): L615–617, Oct. 1980.

Liutkus, J.; Paraszczak, J.; Shaw, J.; Hatzakis, M.
Poly-4-chlorostyrene, a new high contrast negative E-beam resist.
Microcircuit Engineering '82. International Conference on Microlithography, Grenoble, France, 5–8 Oct. 1982, p. 266–269.

Lyman, S.P.; Jackel, J.L.; Liu, P.L.
Lift-off of thick metal layers using multilayer resist.
Journal of Vacuum Science and Technology 19 (4): 1325–1328, Nov/Dec. 1981.

Matsui, S.; Endo, N.
New tri-level structures for submicron photolithography.
International Electron Devices Meeting, Technical Digest, San Francisco, California, 13–15 Dec. 1982, p. 395–398.

Mochiji, K.; Wada, Y.; Obayashi, H.
Improved dry etching resistance of electron-beam resist by ion exposure process.
Journal of the Electrochemical Society 129 (11): 2556–2559, Nov. 1982.

Morita, S.; Tamano, J.; Hattori, S.; Ieda, M.
Plasma polymerized methyl-methacrylate as an electron-beam resist.
Journal of Applied Physics 51 (7): 3938–3941, July 1980.

Namatsu, H.; Ozaki, Y.; Hirata, K.
High resolution trilevel resist.
Journal of Vacuum Science and Technology 21 (2): 672–676, July–Aug. 1982.

Namatsu, H.; Ozaki, Y.; Hirata, K.
Hydrocarbon–oxygen mixture as a resist etching gas with highly anisotropic etching feature.
Journal of the Electrochemical Society 130 (2): 523–525, Feb. 1983.

Nishida, H.; Umezaki, H.; Koyama, N.; Sugita, Y.
New single-mask approach to bubble device fabrication.
IEEE Transactions on Magnetics MAG-19 (1): 2–6, Jan. 1983.

Ogata, N.; Sanui, K.; Azuma, C.; Tanaka, H.; Oguchi, K.; Nakada, T.; Takahashi, Y.
Relationship between electron sensitivity and chemical structure of polymers as EB resists – 1. Electron sensitivity of various polyamides.
Journal of Applied Polymer Science 28 (2): 699–708, Feb. 1983.

Ohnishi, Y.
Poly(vinylnaphthalene) and its derivatives as E-beam negative resists.
Journal of Vacuum Science and Technology 19 (4): 1136–1140, Nov/Dec. 1981.

Randall, J.N.; Wolfe, J.C.
Preparation of X-ray lithography masks using a tungsten reactive ion etching process.
Applied Physics Letters 41 (3): 247–248, Aug. 1, 1982.

Ray, G.W.; Peng, S.; Burriesci, D.; O'Toole, M.M.; En-Den, L.
Spin-on glass as an intermediate layer in a tri-layer resist process.
Journal of the Electrochemical Society 129 (9): 2152–2154, Sept. 1982.

Sato, M.; Kawabuchi, K.; Yoshimi, M.; Yamazaki, T.
Submicron electron-beam patterning of aluminum by a double-layer pattern transfer technique.
Journal of Vacuum Science and Technology 19 (4): 1329–1332, Nov/Dec. 1981.

Shiraishi, H.; Taniguchi, Y.; Horigome, S.; Nonogaki, S.
Iodinated polystyrene: An ion-millable negative resist.
Polymer Science and Engineering 20 (16): 1054–1057, Nov. 1980.

Srinivasan, R.
Action of far-ultraviolet radiation (185 nm) on poly(ethylene terephthalate) films: A method for controlling dry etching.
Polymer 23 (13): 1863–1864, Dec. 1982.

Sukegawa, K.; Sugawara, S.
Negative electron resist with high lithographic performance.
Electrical Communications Laboratory Technical Journal 30 (12): 2943–2950, 1981.

Tracy, C.J.; Mattox, R.
Mask considerations in the plasma etching of aluminum.
Solid State Technology 25 (6): 83–88, June 1982.

Tsuda, M.; Oikawa, S.; Kanai, W.; Yokota, A.; Hijikata, I.; Uehara, A.; Nakane, H.
A principle for the formulation of plasma developable resists and the importance of dry development of submicron lithography.
Journal of Vacuum Science and Technology 19 (2): 259–261, July/Aug. 1981.

Vogel, S.F.
Photolithographic mask structured to remove redeposited material after ion milling or sputter etching.
IBM Technical Disclosure Bulletin 21 (3): 1218–1219, Aug. 1978.

Wada, O.; Yamamoto, S.; Kobayashi, K.; Taguchi, A.; Toyama, Y.
Mask preparation for small dimension ion milling by two step lift-off process.
Journal of the Electrochemical Society 124 (6): 959–960, June 1977.

Wada, Y.; Mochiji, K.; Obayashi, H.
Reactive ion etching resistant negative resists for ion beam lithography.
Journal of the Electrochemical Society 130 (1): 187–190, Jan. 1983.

Wade, T.E.
Low temperature double-exposed polyimide/oxide dielectric for VLSI multilevel metal interconnection.
IEEE Transactions on Components, Hybrids, and Manufacturing Technology CHMT-5 (4): 516–519, 1982.

White, L.K.; Meyerhofer, D.
Positive resist processing considerations for VLSI lithography.
RCA Review 44 (1): 110–134, Mar. 1983.

Wilkins, C.W. Jr.; Reichmanis, E.; Chandross, E.A.
Lithographic evaluation of an o-nitrobenzyl ester based deep UV resist system.
Journal of the Electrochemical Society 129 (11): 2552–2555, Nov. 1982.

Yamazaki, T.; Suzuki, Y.; Uno, J.; Nakata, H.
The role of a photoresist film on reverse gas-plasma etching of chromium films.
Japanese Journal of Applied Physics 19 (7): 1371–1376, 1980.

Yoneda, Y.; Kitamura, K.; Miyagawa, M.
Polydiallylorthophthalate resist for electron-beam lithography.
Fujitsu Scientific and Technical Journal 18 (3): 453–467, Sept. 1982.

Yoshikawa, A.; Ochi, O.; Mizushima, Y.
Dry development of Se–Ge inorganic photoresist.
Applied Physics Letters 36 (1): 107–109, 1 Jan. 1980.

SUBJECT INDEX*

Aluminum (Al)
  plasma-assisted etching of, 3–36
    chlorine-based gases, 5–8, 17–28
    contaminants, 9, 25–28
    edge profile, 20–21, 22f–23f, 29
    etch chemistry, 8–10, 15–33
    etch mechanisms, 5–15
    etch rate, 18–19f
    polymerization, 7–8, 22, 28
    post-etch corrosion, 33–34
    safety, 35–36
    scavenging, 26–28
    selectivity, 29–33
    sidewall etching, 20–24
  reactive ion beam etching (RIBE) of, 163–170
Aluminum alloys
  plasma-assisted etching of, 33
  reactive ion beam etching (RIBE) of, 163–170
Aluminum–copper–silicon (AlCuSi) alloy, 161, 163–170
  etch chemistry, 169–170
  oxygen effects on etch, 166f, 167–168
Aluminum trichloride ($AlCl_3$)
  problems with, 17, 28, 31, 33
Anisotropy *see* Edge profile
Annealing, 68

Barrel reactor, 116–117
  compared to other dry etch techniques, 121t
Batch mode wafer processing, 28–29

Boron trichloride (BCl_3), 8, 18
  as oxygen scavenger, 26–27
  contaminant of GaAs etching, 95
  toxicity, 35–36
Broad-beam ion source *see* Ion sources

Carbides (SiC), RIBE applications, 199–200
Carbon tetrachloride (CCl_4), 18, 22–23
  decomposition products of, 5–7
Chlorine-based gases, in
  aluminum etching, 5–8, 17–28
  metal silicide etching, 60–62
  polycide etching, 65–66
  polysilicon etching, 172–173, 173f
  refractory metal etching, 56–58
  SiO_2 and Si etching, 171–175f, 182–183
Cold-cathode ion source, 146–147
Contaminants in plasma, 25–28
Copper alloy, 33, 34

Dry etching, 3–5
  bibliography, 223–294
  device damage, 208–210
  information sources, 217–222
  techniques compared, 115–121t
  III–V semiconductors, 81–108

Edge profile
  aluminum etching, 20–21, 22f–23f, 29
  ion bombardment, 10–11
  molybdenum etching, 51, 57f–58
  $MoSi_2$ etching, 59–60f
  III–V semiconductors, 90–91f, 102–104
    with RIBE, 126, 191, 200–208

---

* If a page number is followed by an f, the entry refers to the figure(s) on that page, similarly a number followed by a t refers to a table.

Edge profile (cont'd)
  tungsten, 51
Electron cyclotron resonance (ECR), 143–144, 145f
End-point detection, 68–69, 130
  bibliography, 283–288
Etch chemistry, of, 8t
  aluminum, 8–10, 15–33
    etch product volatility, 9–10
    initiation period, 15–16
      effect of H_2O on, 26f
    oxygen as a contaminant, 24–28
    polymerization, 7–8, 22, 28
    post-etch corrosion, 33–34
    safety, 35–36
    selectivity, 29–33
    sidewall etching, 20–24
    scavenging, 26–28
  AlCuSi, 169–170
  aluminum trichloride ($AlCl_3$) problems, 17, 28, 31, 33
  electrode materials, 7–8
Etch mechanisms, in
  aluminum etching, 5–15
  refractory metal/metal silicide etching, 70–73
  RIBE, 147–162
    gas-surface chemistry
      inert ion beam, 149–155
      neutral species, 160–162
      reactive ion beam, 155–160
    III–V semiconductors, 94–107

Fluorine-based gases, in
  aluminum etching, 9–10
  metal silicide etching, 59–60
  polycide etching, 64–65
  polysilicon etching, 176–182
  refractory metal etching, 51–55f
  SiO_2 and Si etching, 176–182, 183–185

Gallium arsenide (GaAs), 88, 97t–98t
  BCl_3 contamination, 95
  edge profile, 103–104
  field effect transistor (FET), 83f
  native oxide, 105

GaAs (cont'd)
  oxygen concentration, 106
  product desorption, 98, 101–102
  RIBE applications, 202–208
  substrate temperature, effects of, 105
  via hole etching, 85
Gas–surface chemistry
  inert ion beam, 149–155
  neutral species, 160–162
    chemically reactive, 161–162
  reactive ion beam, 155–160
    with reactive gas, 158–160
Gasification reaction, 10–11
Gates, 41–75
  composite, 63–67
    metal-nitride/metal, 67
  single-level, 50–63
    halide-based etching, 58
    metal silicides, 58–63
    refractory metals, 50–58
  structures, 49f

Halide-based gas mixtures, in
  metal silicide etching, 62–63
  polycide etching, 66
  refractory metal etching, 58

Indium phosphide (InP), 98t
  edge profile, 103–104
  product desorption, 98–102
  RIBE application, 200–201
  substrate temperature, effects of, 105
Indium gallium arsenide (InGaAs), 84f, 86–87
Interconnect layer, 41–42
  high and low temperature materials, 43t, 44t
Ion-assisted chemical etch
  inert, 123–124f, 126, 139
    mechanisms, 150–155
  reactive, 125f, 126
    mechanisms, 155–160
    with reactive gas, 158–160
Ion beam milling, 4, 119–121
  bibliography, 272–283
  compared to other dry etch techniques, 121t

Subject Index

Ion bombardment, 10–15
  edge profile, 10–11
  energy, 14–15, 19f
  momentum transfer, 30–31
  self bias, 14
Ion-enhanced etching, of
  III–V semiconductors, 102–104
Ion sources, 130–147
  broad-beam (Kaufman-type), 122f, 130–139
    electric potential, 134f
    hollow cathode, 136–138
    ion extraction, 133–134
    magnetic fields, 139–142
    use of reactive gases, 135–139
      hot-filament problems, 135–136
    low energy, 142–143
  broad-beam (non-Kaufman-type), 143–147
    cold-cathode, 146–147
    electron cyclotron resonance (ECR), 143–144, 145f
    magnetron ion guns (MIG), 144–146

Josephson junctions, 51

Kaufman-type ion source, 130–135
  using reactive gases, 135–139
  see also Ion sources

Magnetron ion guns (MIG), 144–146
Masking materials
  bibliography, 288–294
Metal nitride, 63, 67
Metal-oxide-semiconductor (MOS) gates, 41–75
Metal silicides, 43–45, 58–63
  edge profile, 60f
  etch gases, 47t
  etch mechanisms, 70–73
  etch rates, 59f, 61f–62f
  etching, 45–50
  gates, 49f, 58–63
    chlorinated, 60–63
    fluorinated, 59–60, 62–63
  polycide, 44–45, 50, 63–66, 68
  properties of, 44f

Metal silicides (cont'd)
  RIBE applications, 192–199
Mirror facet formation, 89–92
Molybdenum silicide ($MoSi_2$), 59–63
  RIBE applications, 192–196
Momentum transfer, 30–31

Native oxide, on
  aluminum, 15, 17–18
  III–V compounds, 105
Neutral species, 148, 160–162
  chemically reactive, 161–162

Oxygen (O_2)
  as additive to plasma gases, 45, 71–73
  concentrations in GaAs plasma etching, 106
  as contaminant in Al etching, 24–28
  effect on etch rates of
    AlCuSi, 166f, 167–168
    Mo, 56f–57, 67
    $MoSi_2$, 62f
    $TaSi_2$, 67, 68
    tungsten, 55f
  effect on reactivity, 95–96
  photoresist etch gas, 185–190f

Parallel plate etch systems, 11–14, 48f, 117
  compared to other dry etch techniques, 121t
  glow discharge, 11f
  plasma sheath region, 12f
  potential in
    nonsymmetrical parallel plate, 14f
    symmetrical parallel plate, 13f
Parallel-plate reactor *see* Parallel plate etch systems
Photoelectrochemical etching, 88, 89
Photoresist, 57, 185–192
  in Ar/O_2, 185–190f
  temperature problems, 192
Planar reactor *see* Parallel plate etch systems
Plasma-assisted etching, of
  aluminum and Al alloys, 3–36
Plasma bridge neutralizer, 136
Plasma contaminants, in aluminum etching, 25–28

Plasma diagnostics
  bibliography, 283–288
Plasma etching, 116–117f
  bibliography, 228–254
  characteristics, 45–50
  mechanisms, of
    refractory metals/metal silicides, 70–73
    III–V semiconductors, 94–107
    refractory gates, 41–75
    techniques, 45–46, 48f
  models, 8–10
Polycide, 44–45, 50, 63–66, 68, 192, 194–196
  chlorinated plasmas, 65–66
  fluorinated plasmas, 64–65
  halide-based gas mixtures, 66
  RIBE applications, 192, 194–196
Polymerization, 7–8, 22, 28, 138
Polysilicon (poly-Si), 41, 43–45, 63–66
  in RIBE, 172–173, 176–182
Product desorption, of
  III–V semiconductors, 96–102, 103

Reactants, in III–V semiconductor etching, 106–107
Reactive ion beam etching (RIBE), 115–211
  bibliography, 272–283
  characteristics, 121–126, 147–149, 210–211
  compared to other dry etch techniques, 121t
  damage, 208–210
  development, 120, 210–211
  equipment design, 126–130, 210t
  ion sources, 130–147
  materials, 163–208
    aluminum and Al alloys, 163–170
    new material applications, 192–208
    photoresist and other organics, 185–192
    polysilicon, 172–173, 176–182
    SiO_2 and Si, 171–185
Reactive ion etching (RIE), 48f, 118–119
  bibliography, 254–272
  compared to other dry etch techniques, 121t
  edge profile, 20–21, 22f, 23f, 90–92, 104
  mirror facet formation, 90–92

Refractory metals, 43–45
  edge profile, 55f, 57f
  etch gases, 46t
  etch mechanisms, 70–73
  etch rate, 52f–55f, 56f
  etching, 45–50
  gate, 49f, 50–58
    chlorine-based gases, 56–58
    fluorine-based gases, 51–55f
    halide-based gas mixtures, 58
  materials, 50–51
  properties
    effect on etching, 67–68
    of chlorides/oxychlorides, 73t, 74t–75t
    of fluorides/oxyfluorides, 70t, 72t
  RIBE, applications, 198f–199
Resists
  bibliography, 288–294
  material problems, 30–33

Safety in plasma etching, 35–36
Scavenging, 26–28
Schottky barriers, 93
Selectivity, 29–33
Self bias, 14
Sheet resistivity, 42, 68
Sidewall etching, 20–24
Sidewall profile *see* Edge profile
Silicon dioxide (SiO_2) and Si, 171–185
  etch gases
    chlorine-based, 171–175f
    fluorine-based, 176–182
  mechanisms, 182–185
    chlorine-based chemistry, 182–183
    fluorine-based chemistry, 183–185
  summary table, 171t
Silicon tetrachloride ($SiCl_4$), 24–25, 27–28
Single wafer etching (SWE) mode, 29
Sputtering, 4–5, 48f
  reactive, 155–157
  physical, 157
Surface morphology of substrate, 105
Synergism, ions and neutral reactants, 95, 103

Tantalum silicide (TaSi$_2$), 64–65, 67, 68
  RIBE applications, 196–197f
Temperature, effects of, 30–32, 87–88, 90
  on substrate, 67, 102–103, 105, 107–108
Thermionic cathodes (hot filament),
  problems with reactive gases in RIBE,
  135–139
III–V semiconductors, 81–108
  applications, 82t–93
    electronic, 92–93
    macroscopic, 82–89
      chip separation, 86–88
      other applications, 88–89
      via hole, 85
    optical, 89–92
    using RIBE, 200–208
  etch gases, 95–96
  etch mechanisms, 94–107
    ions and neutral reactants, 95
    ion-enhanced etching, 102–104

III–V semiconductors (cont'd)
  etch mechanisms (cont'd)
    general concepts, 94–96
    product desorption, 96–102
    reactants, 106–107
    surface morphology, 105
  etch products, halides, 97t
  summary, 97t–98t, 107–108

Very-large-scale-integrated (VLSI) circuits
  etching, 45
  interconnects, 41–42
  metallization, 3, 43–44

Wafer processing
  batch mode, 28–29
  single wafer etching (SWE) mode, 29
Water (H$_2$O)
  as a contaminant, 25–28
  effect on initiation period, 26f
Wet chemical etching, 86, 88, 90, 92, 93, 115

Tantalum silicide (TaSi$_2$), 64–65, 67, 68
  RIBE applications, 196–197f
Temperature, effects of, 30–32, 87–88, 90
  on substrate, 67, 102–103, 105, 107–108
Thermionic cathodes (hot filament),
  problems with reactive gases in RIBE, 135–139
III–V semiconductors, 81–108
  applications, 82t–93
    electronic, 92–93
    macroscopic, 82–89
      chip separation, 86–88
      other applications, 88–89
      via hole, 85
    optical, 89–92
    using RIBE, 200–208
  etch gases, 95–96
  etch mechanisms, 94–107
    ions and neutral reactants, 95
    ion-enhanced etching, 102–104

III–V semiconductors (cont'd)
  etch mechanisms (cont'd)
    general concepts, 94–96
    product desorption, 96–102
    reactants, 106–107
    surface morphology, 105
  etch products, halides, 97t
  summary, 97t–98t, 107–108

Very-large-scale-integrated (VLSI) circuits
  etching, 45
  interconnects, 41–42
  metallization, 3, 43–44

Wafer processing
  batch mode, 28–29
  single wafer etching (SWE) mode, 29
Water (H$_2$O)
  as a contaminant, 25–28
  effect on initiation period, 26f
Wet chemical etching, 86, 88, 90, 92, 93, 115